THE MISSOURI

Books by Stanley Vestal:

KIT CARSON
SITTING BULL
WARPATH
KING OF THE FUR TRADERS
BIGFOOT WALLACE
THE OLD SANTA FE TRAIL
NEW SOURCES OF INDIAN HISTORY
MOUNTAIN MEN
SHORT GRASS COUNTRY
FANDANGO, BALLADS OF THE OLD WEST
'DOBE WALLS
REVOLT ON THE BORDER
HAPPY HUNTING GROUNDS
THE MISSOURI

Rivers of America books already published are:

KENNEBEC by Robert P. Tristram Coffin
UPPER MISSISSIPPI by Walter Havighurst (New Revised Edition, 1944.)
SUWANNEE RIVER by Cecile Hulse Matschat
POWDER RIVER by Struthers Burt
THE JAMES by Blair Niles
THE HUDSON by Carl Carmer
THE SACRAMENTO by Julian Dana
THE WABASH by William E. Wilson
THE ARKANSAS by Clyde Brion Davis
THE DELAWARE by Harry Emerson Wildes
THE ILLINOIS by James Gray
THE KAW by Floyd Benjamin Streeter
THE BRANDYWINE by Henry Seidel Canby
THE CHARLES by Arthur Bernon Tourtellot
THE KENTUCKY by T. D. Clark
THE SANGAMON by Edgar Lee Masters
THE ALLEGHENY by Frederick Way, Jr.
THE WISCONSIN by August Derleth
LOWER MISSISSIPPI by Hodding Carter
THE ST. LAWRENCE by Henry Beston
THE CHICAGO by Harry Hansen
TWIN RIVERS: The Raritan and the Passaic, by Harry Emerson Wildes
THE HUMBOLDT by Dale Morgan
THE ST. JOHN'S by Branch Cabell and A. J. Hanna
RIVERS OF THE EASTERN SHORE by Hulbert Footner
THE MISSOURI by Stanley Vestal

SONGS OF THE RIVERS OF AMERICA edited by Carl Carmer

THE RIVERS OF AMERICA

Edited by
HERVEY ALLEN

Associate Editor JEAN CRAWFORD
As Planned and Started by
CONSTANCE LINDSAY SKINNER
Art Editor FAITH BALL

THE
MISSOURI

by STANLEY VESTAL

Illustrated by
GETLAR SMITH

Maps by GEORGE ANNAND

FARRAR & RINEHART
INCORPORATED
New York　　　　　　　Toronto

This book has been manufactured in accordance with paper conservation orders of the War Production Board

COPYRIGHT, 1945, BY WALTER STANLEY CAMPBELL
PRINTED IN THE UNITED STATES OF AMERICA
BY J. J. LITTLE AND IVES COMPANY, NEW YORK
ALL RIGHTS RESERVED

To
AUNT HARVIE
in gratitude, admiration,
and love

Contents

RECONNAISSANCE 5

HIGHWAY

I. THE FATHER OF NAVIGATION . . . 11
II. MISSOURI MARATHON 18
III. STEAMBOAT 'ROUND THE BEND . . . 34
IV. LONG'S DRAGON 50
V. PRINCESS OF THE MISSOURI 59
VI. THE BIG MUDDY 66
VII. BEAVER TAILS AND WILD HONEY . . . 73

BOUNDARY

VIII. THE MORMON MIGRATION 85
IX. DIVING FOR FURS 98
X. HELL ON THE BORDER 106
XI. KAYCEE 123
XII. AK-SAR-BEN 130
XIII. CHIEF BLACKBIRD AND SERGEANT FLOYD . 139
XIV. THE PETRIFIED MAN 149
XV. RANGE AND GRANGE 162

CONTENTS

OUTPOST

XVI.	MOUNTAIN MEN	183
XVII.	STANDING ROCK	196
XVIII.	FOUR BEARS	204
XIX.	CUSTER AND COMANCHE	219
XX.	GHOST DANCE	226
XXI.	THE MISSOURI RIVER WOMEN	243
XXII.	THE LITTLE MISSOURI	259
XXIII.	FORT UNION	267
XXIV.	THE GREAT DAM	278
XXV.	HONEYMOON CRUISE	285
XXVI.	WHITE CASTLES	292
XXVII.	CHIEF JOSEPH'S LAST BATTLE	299
XXVIII.	GYPSIES OF THE UPPER RIVER	311
XXIX.	GREAT FALLS	316
XXX.	THE GATES OF THE MOUNTAINS	322
XXXI.	THREE FORKS	330
NOTES		335
ACKNOWLEDGMENTS		345
BIBLIOGRAPHY		349
INDEX		355

Reconnaissance

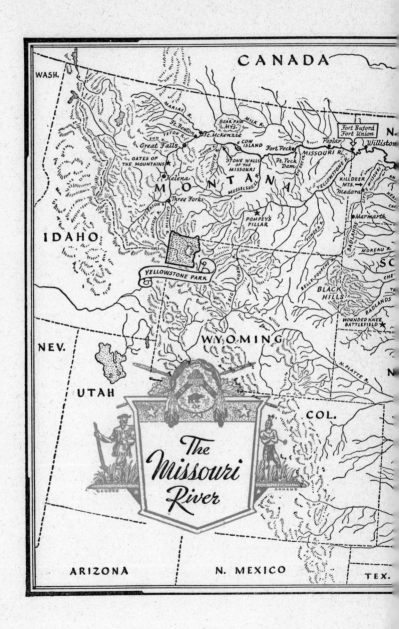

Reconnaissance

THERE are rivers that exist in time—or in eternity. But the Missouri River quite manifestly exists in space. It is on its way. It moves magnificently over vast distances. It is forever going places. Not only that, it is forever doing things.

There are streams that have no story except that of the people on their banks, but the Missouri River is a story in itself—and no idyll or eclogue either, but an heroic poem, an epic. It is a thoroughly masculine river, a burly, husky bulldozer of a stream, which has taken on the biggest job of moving dirt in North America. It has been well-named the Big Muddy.

For almost a century the Missouri was a principal highway of the biggest business in North America—the fur trade. But the traffic in furs was only a small portion of the business handled on the stream. We hear much of the Santa Fe and Oregon trails and other famous roads leading into and across the Great American Desert, and forget that nearly all travelers on those overland trails began their journey with a boat trip on the Missouri River. Thousands of the wagons which rolled to Santa Fe and Oregon and California traveled on boats from St. Louis to Kansas City or St. Joseph before their iron tires could bite into the grass of the Great Plains.

But the Missouri River was not only the main highway to the West, it was also the base of operations for the winning of the West—an unbroken chain of forts and camps, missions, trading posts and Indian agencies, all devoted to that conquest.

Throughout much of its long course, also, it was a definite boundary or barrier between two kinds of country, two climates, two cultures, two ways of life. On its left bank lay beautiful woodlands, lakes, and prairies; from its right bank rose the magnificent Plains, stretching away to the Shining Mountains, the backbone of the continent.

On the east bank the first explorers found Indian woodsmen living in smoky wigwams, skulking among the trees, stalking deer afoot in soft-soled moccasins and garments decorated with flowing flower patterns—Indians whose warfare consisted of sniping, ambush, and massacre. On the west bank they found hardy buffalo hunters and warriors on horseback who lived in stately tipis decorated with bold geometric patterns—men of dash and spirit, who raided in the open and looked for no better cover than a rawhide shield or the body of a racing pony, no better protector than a good right arm.

Since white men occupied the valley, the river is still the boundary between two cultures. On the east bank you find the plow, the farm, the barnyard, the tall corn, and the rifle. On the west bank you have the saddle, the ranch, the corral, the beef steer, and the six-gun. If anyone wishes to locate the line "where the West begins" he should look for it in the middle of the Missouri River.

It divides or washes the boundaries of seven states—Missouri, Kansas, Nebraska, Iowa, South Dakota, North Dakota, Montana; four state capitals stand upon its banks.

Yet the Missouri was first of all a *highway*, and it is as a highway that it has captured the imagination of mankind—a perilous trail leading from Mississippi swamps to the snow peaks of the Rockies; a trail passing from the warm South to a country bleak and cold and bare as Siberia.

Pioneers who followed that trail can scarcely have foreseen the railroads and pavements and airways which have superseded the boats they manned. We are in danger of forgetting the lure of that great river.

Yet we can travel with them, in imagination, and so share their feelings and sensations, their fears and hopes, the fun and hardships and perils which they knew. That experience gave them a certain quality, a habit of mind that has produced ways of life that differ somewhat from those of other regions. Who were these people? Where did they come from? What did they look for—and find? How did they get that way?

To understand the people on the Missouri, it is first necessary to know the river itself; for it was always an active member of the community—though not always one in good standing. The pioneers lived with one foot in the Big Muddy and it had a profound effect upon their character, feelings, behavior, and ideas—an effect which will gradually become clear as our story grows. So, whenever we speak of the river, we are actually talking about the folks who had to live with it. For the Missouri is not just old man river—he is *the* Old Man. And the folks we are to meet are his folks.

For them that river, that highway, meant cold and wet and violent storms, beating hail and caving banks, snags and rapids and wrecked boats, an avalanche of high water sweeping all before it or a narrow channel under the bank where the skulking Sioux lurked in the willows with an arrow on his bowstring. To them it meant terror and gaiety, sometimes famine and pestilence, great fortunes or cruel wars; and so, if we are to know their river, we must live with it as they did. We can recapture their experience, for they have left us their diaries, their newspapers, their letters, their books—and their descendants. The river itself is still there and going strong.

Let us start at the mouth as most of them did, and work

our way up those 2,500 stubborn, watery miles, seeing what they saw, feeling what they felt. The trip should be rewarding.

For as George Fitch has so well said, "There is only one river that goes traveling sidewise, that interferes in politics, rearranges geography and dabbles in real estate; a river that plays hide-and-seek with you today, and tomorrow follows you around like a pet dog with a dynamite cracker tied to his tail. That river is the Missouri."

All aboard! In the words of the old song:

We are bound for the wild Missouri!

Highway

CHAPTER I

The Father of Navigation

THE trouble with going up the Missouri River in a boat is that you have to take the boat along.

Now the Missouri is about equally unsuited to every kind of river craft: too swift for oars, too deep for poles, too crooked for sails, too shallow for keels, and without permanent banks for a towpath. In fact, until Manuel Lisa came along, nobody supposed that it could be navigated with any success, or for any distance, or by anything bigger than a canoe.

The first white man to see the mouth of the Missouri and to leave a record of it was Father Marquette, who reached the mouth in June, 1673. Some Indian had given him a name for the stream, Pekitanoui. Paddling in his frail birchbark canoe down the Mississippi, the good father was appalled by what he saw. He writes:

"As we were gently sailing down the still clear water, we heard a noise of a rapid into which we were about to fall. I have seen nothing more frightful, a mass of large trees entire with branches, real floating islands came from Pekitanoui, so impetuous that we could not without great danger expose ourselves to pass across. The agitation was so great that the water was all muddy, and could not get clear."[1]

Ever since Father Marquette first saw that wild and wicked river, boatmen on the stream have felt about it just as he did. Probably there has been more violent and heartfelt cursing on that river than on any highway of its length on this continent.

No other river was ever so dead-set against being navigated. No other river was ever so determined on having its own way. It is as ruthless as it is unpredictable, a regular Caliban among rivers.

On that stream, the inevitable never happens. It is ready to try anything once. It is an honest-to-God American river —*the* American river.

At that, it is now a mere trickle to what it must have been at the end of the ice age, when the mighty glaciers were melting on its headwaters. Human imagination staggers to consider what the Missouri was up to then.

Originating in the union of three sizable streams, the Missouri was born self-sufficient, a Hercules bursting from its cradle in the mountains. In prehistoric days, its headwaters flowed clear and vigorous north and east, heading toward Hudson's Bay or the basin where the Great Lakes lie today. But the coming of the icecap changed all that, and turned the stream south between the glacial drift and the alluvial apron of the Rockies. The river began as a hopeful, pushing, enterprising stream; but after it joined forces with the turbid Yellowstone, it seems to have lost its clear purpose, though increasing its original force.

Ever since the glaciers turned it south, the frustrated stream has been restless, seeking its lost channel, never content with the bed it occupies. A book could be filled with its unaccountable antics, which continue to this day:

"To look at it, some people would think it was just a plain river running along in its bed at the same speed; but it ain't. The river runs crooked through the valley; and just the same way the channel runs crooked through the river. The river changes whenever it feels like it in the valley; and just the same the channel changes whenever it feels like it in the river. The crookedness you can see ain't half the crookedness there is." [2]

Cow Island was once on the Kansas side of the channel. Charles Keane had a saloon there, selling liquor legally to all comers. Suddenly, in 1881, the great flood shifted the channel from the Missouri to the Kansas side, and Keane was arrested on a charge of selling intoxicating liquor, haled before the circuit court in Platte City, Missouri, and promptly convicted. The case was then carried to the court of appeals in Kansas City.

That court held that *if* the state boundary (defined as the center of the navigable channel, "the river as it runs") had changed gradually and imperceptibly, the boundary would have changed with it. *But,* since there had been a sudden avulsion, the river leaving its old bed dry and seeking a new course, the boundary remained where it had been before the sudden shifting of the channel occurred. Accordingly, the court of appeals reversed the judgment of the circuit court, which had no jurisdiction, as the saloon was still in Kansas. This verdict was hailed with joy by thirsty Missourians, who no longer had to cross the river to get a drink. . . .

"Ask the citizen of a Missouri River town on which side of the river he lives, and he will look worried and will say: 'On the east side when I came away.' Then he will go home to look the matter up and, like as not, will find the river on the other side of his humble home and a government steamboat pulling snags out of his erstwhile cabbage patch.

"It makes farming as fascinating as gambling too. You never know whether you are going to harvest corn or catfish. . . .

"It is the hungriest river ever created. It is eating all the time—eating yellow clay banks and cornfields, eighty acres at a mouthful; winding up its banquet with a truck garden and picking its teeth with the timbers of a big red barn. Its yearly menu is ten thousand acres of good, rich farming land, several miles of railroad, a few hundred houses, a forest or two and uncounted miles of sandbars." [3]

Engineers are a busy lot of men on the Missouri. One of them built a bridge. He says, "I spent a year putting the bridge over the river; I've spent my time ever since, keeping the river under the bridge."

"Of all the variable things in creation the most uncertain are the action of a jury, the state of a woman's mind, and the condition of the Missouri River." [4]

The first voyageurs soon found out how different the Missouri was from other streams in North America. They discovered that paddling up that stream in a thin-skinned birchbark canoe was like trying to kill a porcupine with a toy balloon. They learned that the Indians along its banks had a better notion. The Indians used pirogues—dugout canoes. Indeed there are those who insist that the name of the river means Wooden Canoe.

The dugout was made, as its name suggests, by gouging out or burning out the inside of a log into a thin wooden shell. The beam of the vessel would necessarily be less than the thickness of the log, and its length would depend upon the length of the log—or the wish of the man who made it. In early times trees on the Lower Missouri were sometimes very large. One traveler writes that he had never seen anything comparable in the British Isles. Cottonwoods and walnut trees sometimes measured six feet in diameter and might extend forty feet up to the first limb.

These dugout canoes sometimes had a square, undercut stern without a rudder, sometimes a shallow keel, and a pointed or rounded bow with a projection at the top in front through which a hole was bored or burned to receive the mooring rope. The makers sometimes left bulkheads at intervals to strengthen the boat and to provide compartments in which to stow cargo.[5]

The average dugout had a beam of two or three feet and was long enough for a crew of three: one to paddle forward, one amidships, and a steersman in the stern. No doubt some of them were smaller and a few much larger. Rudolph Friederich Kurz, the artist, made a sketch of such an Oto pirogue not more than eight feet long, which could hardly have carried more than two people. In that charming painting by George Caleb Bingham, made about 1845, "Fur Traders Descending the Missouri," [6] we see a dugout about twelve feet long with a prow projecting above the gunwale eight or ten inches, and the cargo stowed amidships. The crew consists of two men and their black cat. Probably Indians on the Lower River had war canoes much larger than these.

Such a wooden canoe had solid advantages. Though heavy and sometimes awkward to handle, it was practically unsinkable and sturdy enough to endure some contact with snags and gravel. Still, it could be upset, and the first improvement the voyageurs made was intended to prevent that disaster. They fastened two canoes side by side at a distance of eight or ten feet apart and then decked over the intervening space. Such a vessel could carry a considerable cargo, depending on its size, and was not easily overset. Much of the earliest traffic on the river was shipped on such craft. Lewis and Clark called them "rafts."

Sawmills were slow in coming to the Missouri Valley and therefore flatboats like those commonly used on the Mississippi were at first not seen there. The river was too turbulent and too swift for such unwieldy craft, and by the time the settlers had the planks to make flatboats, the boatmen had adopted a vessel better able to meet conditions on that stream. This was the Mackinaw—a flat-bottomed boat of light draft with a sharp prow and a rudder, unable to buck the strong currents going up, but able to avoid obstacles coming down. The Mackinaw carried much of the downriver traffic even after steamboats came in.

Failing a Mackinaw or a pirogue, the skin boat was used. This was a framework of wood covered with buffalo hides tightly sewed together with sinew, the seams being plastered with gumbo mud or buffalo tallow to keep the sinew dry. Though light and speedy, these boats did not stand up in service, and soon rotted through. They leaked too, buffalo hide being so porous, but were good for a short, swift trip.

Men who traveled in any of these boats spent half their time in the water. Whenever she ran aground on a bar and threatened to capsize, everybody piled overboard and held her right-side-up until the sand below had washed away. It may be said of the boatmen of such small craft that they traveled *with* a boat rather than aboard it. It was no wonder that men declared the river unnavigable.

Manuel Lisa, a swarthy Spanish fur trader, first called the tune for navigating the waters of the Missouri River in 1811, and his music was in jig time. While other men dawdled and drifted and pottered about the stream with their nondescript craft, he squared away and made a voyage. Lisa had been a sea captain; he could not be content with dugouts and rafts. He wanted a craft with a keel, a deck, a cabin, and a sail. He had a keelboat built to order especially for the Missouri, and quickly demonstrated what he could do with it.

The season of high water, of the spring freshet, on the river is short—from mid-May to about July 20th. Anyone who hoped to go up the river in a sizable boat and return the same year had to make the round trip within that short season. Once for all, Lisa showed how to go about it. Speed was his slogan. He is rightly called the Father of Navigation on the Missouri.

Such a title implies a great deal. They say that Missouri mules get their cussedness from drinking the waters of the

Missouri River, but the cussedness of a mule is small potatoes compared to the cussedness of the Big Muddy. Nowadays few people have any notion what Manuel Lisa and his fellow navigators had to contend with.

He ran a great race for high stakes.

CHAPTER II

Missouri Marathon

A RIVER is manifestly made for a boat race, and the Missouri has many tales of races between steamboats, bullboats, canoes, and even ferries. But the most sensational and the best known on the Big Muddy was that great race between keelboats in the spring of 1811.

That was no mere contest for a silver cup. The glittering prize was monopoly of trade with the Sioux. The forfeit seemed likely to be the scalps of the losers.

The two crews belonged to two rival fur companies—the Astorians and the Missouri Fur Company. The course was upstream from the mouth of the river nearly 1,200 miles. Much of the way the river passed through the country of hostile Indians.

It was a grudge race. From the start the two crews were at daggers' points, each one believing it had been bitterly wronged by the other. Each party, moreover, believed that the other would arm the hostile Sioux to destroy its rival. It was a plain case of devil take the hindmost. Some time before, two Astorians had been held prisoner and then driven out by the Sioux—as they claimed, at the instigation of Manuel Lisa. Lisa, on his part, blamed the Astorians because the Indians had put his partner to flight and had broken up the business of his Missouri Fur Company on the Upper River. Manuel Lisa and Wilson P. Hunt, who commanded the rival party, trusted each other about as far as one of them could throw a 70-foot keelboat.

Hunt had a flotilla of four boats, one of twenty tons burden, and each manned by picked boatmen from French Canada—indefatigable, singing voyageurs, whose skill and daring on the St. Lawrence were known to everyone. Hunt also had able lieutenants, enemies of Lisa; and Pierre Dorion, interpreter, who had contracted to work for Lisa but had been lured away by Hunt. These men knew the Missouri like the palms of their hands.

Hunt had wintered some three hundred miles up the river, and had come down in the spring with one boat to get the mail. While in the settlements, he did all he could to make it impossible for Lisa to go upriver.

Because of the trouble with the Indians, the Missouri Fur Company had suffered heavy losses. Twenty of their employees had been killed, and they had garnered hardly any furs during the previous season. Lisa's creditors made trouble and caused delays, and even his associates in the company distrusted him.

Lisa had but one keelboat with a green crew of twenty French Creoles from St. Louis, few of whom could boast that they had gone to the Upper Missouri before. It was impossible to induce Anglo-Americans to perform the heavy labor and endure the strict discipline necessary on such a voyage. The captain had one passenger, H. M. Brackenridge. Brackenridge kept a journal of the trip.

With Hunt was a scientist from Scotland, John Bradbury, who also kept a diary. Thus each party made a day-by-day record of its progress—records later published abroad.

A race up the Missouri was inevitably an obstacle race. Lisa was under a heavy handicap. Hunt left St. Charles March 14th, but Lisa, unable to escape the clutches of his creditors, could not get away until April 2nd. Thus he had to make up nineteen days' travel to overtake his rival.

That was a tough assignment.

If you look at the map, you observe that the valley of the river runs north, east, or south, and you might suppose that the stream flowed in those directions only. But the Missouri would surprise you, for very often its current flows west. Sometimes it even flows upstream.

Lisa soon discovered that the water in the river was always rising or falling. When it rose, it lifted all the logs and driftwood lying on a thousand bars and hurled them violently downstream—an avalanche of battering-rams—adding to these headlong missiles by snatching down living trees from its caving banks, to send them with all their boughs and vines and roots whirling and tumbling in the turbulent flood. Along with these went bushes, stumps, and dead logs that had fallen in the forest.

When the water was falling, he found these logs lodged on the bars or accumulated in great rafts across the channel. There they quickly collected sand and were soon matted together by the roots of willows springing hastily up. Such rafts and islands could be formed almost overnight.

Any ordinary stream might be content with mere inert obstacles to navigation, but the Missouri never does things by halves. It set boat traps for Lisa's men: logs which, wedged in the muddy bottom, projected upward, leaning downstream, sometimes reaching above the water, sometimes hidden beneath its turbid surface. Any one of these snags, if struck head-on by a boat, stopped it with a jolt which threw all the crew flat on the deck—if it did not punch a hole in the bottom or tear off the gunwale.

But of course the Missouri could never be content with mere stationary snags—even a million snags. So it anchored the ends of some of the logs to the bottom by their branches or roots, but still allowed them freedom of movement. Borne down by the current, they regularly surged upward again, rising and falling with a violent vibration that earned them the name of "sawyers." When one of these came up under a boat, it meant disaster. If the boat were small, the sawyer turned it over; if large, the sawyer ripped its bottom out.

And then there were the quicksands.

Moreover, the river was never twice the same. Its channel changed overnight. Its depth changed. Where there was a bar one season, there would be a deep eddy the next. In most rivers, the water moves while the bottom lies still: in the Missouri, the bottom sometimes moves as fast as the water—and by no means always in the same direction.

But the river was never content merely to change its snags and sawyers, its bars and banks, its currents and channels. It even moved the rocks around to pester you. A rapid that was clear of rocks one season would be studded with them the next. For, with the help of its ice, the river picked up the rocks and moved them hither and yon every spring. That, you may think, would be enough; but that is because you underestimate the deviltry of the stream. Often it floated rocks downstream in the summertime.

You could see them coming, buoyant as corks—a sight to make your eyes bulge out until you learned that the stones were pumice. And if any of the rocks were too big for the river to move, the river would move itself, and so put the rocks where they would cause the most trouble.

A man is not tempted to sit on the bank and contemplate the Missouri River, since there is a very good chance that the sandy bank will suddenly cave into the water below and end his philosophy in a cold plunge. No, a man navigates that river, and upstream at that. That river was, from the beginning, a road to half America—indeed almost a one-way street. For every man who went down that stream, scores and hundreds went up. It could occur only to a dreamer to go drifting down that watercourse. Everyone else, on beholding its restless energy, immediately feels challenged to force a way up it. So it was with Lisa.

He knew what lay ahead; he had been up the Missouri time and time again. But he had enough of the old Harry in him to go right up again—now that he had his keelboat. His slogan was: "Set poles for the mouth of Yellowstone!"

There probably never was any harder work in this world than getting a keelboat up the Missouri, and to get it up and gain nineteen days on a rival who knew he was pursued and was making all the time he could was a heartbreaking job. Added to these cruel hardships was the probability that the Indians, prompted by Hunt, would lie in wait for Lisa's small party.

On the first of April, Lisa's crew caroused in the taverns of St. Charles, and the tavernkeepers, knowing that Lisa would have to pay before his men could leave, served them liquor lavishly on credit.

Lisa was in no position to be tough with his crew. He let them "take their swing" and put his money on the barrelhead. It was impossible to round them up and get them aboard until two o'clock next day. When all the drunken hands were at last on board, it was too late to go far—what with the high water and swift current. All Lisa hoped for was to get his men out of town and into camp far enough away so that they would not be tempted to walk back to the taverns and start all over again. They camped six miles above St. Charles.

Lisa spared no pains to win that race. He had a fine boat—"The best that ever went up the Missouri." It drew only a little water, was under eighty feet long, and had a cabin or cargo box on deck in which a man could stand upright. The cabin was not so wide as the boat and so left a narrow runway on each side on which the crew worked when using their poles. Forward of the cabin Lisa had rigged a stout mast and topmast with a square mainsail and topsail. The long towrope, or cordelle, was fastened to the top of the mast ready for use. There were oars forward of the cabin, and a great sweep at the stern for steering.

The boat was well-armed. Lisa had a swivel on the bow and two blunderbusses in the cabin—one over his berth and the other over that of Brackenridge. His trade goods, consisting of the usual beads, guns, rifles, powder, lead, tobacco, strouding, awls, kettles, and blankets, were stored compactly in a false cabin built to fool Indians in case they should board the boat. To lighten the craft, the cabin had been built rather flimsily of shingles.

The hardships of the trip began at once. That first night, after the crew had eaten their pot of mush with a pound of tallow in it, and had rolled up in their blankets on the bank, a heavy, cold rain set in and drenched them all thoroughly before daybreak.

Their breakfast consisted of hominy. At noon they were served hard bread or biscuits with a slice of sowbelly. These rations were much better than those usually served to keelboat men on the river in 1811, but Lisa could ill afford to starve his crew. He was proud to have spared no expense. For those days his rations were liberal.

Cold and wet, with hangovers and grave doubts as to whether they would ever come back from this trip, the men clambered aboard and shoved off. But the current soon proved too swift for their oars. Then the men divided into two teams, one on each side of the boat. They seized their long poles shod with iron, ran to the bow, dropped their poles into the water alongside, and setting their shoulders to the knob on the end of the pole, advanced toward the stern, leaning and pushing with all their might and main, trying to shove that boat against the current.

But no matter how much weight and sweat they threw into their efforts toward the stern, while they were running back to the bow, the boat would lose whatever way they had gained. It was impossible to force it against the high water. They had to rig the towing line.

So, overboard they went into icy water up to their shoulders and dragged the boat forward over the bars and shallows in spite of hell and high water. Cordelling was as hard work as ever men did. All day they marched, dripping wet, through water of uncertain depths, plodding through the heavy sand of the bars, their moccasins slipping on the slimy, sloping banks as they went forward, stumbling over fallen trees, clinging to branches and brush hanging from the banks ("bushwhacking," as they called it), unable to find good footing either ashore or in the water.

In the channel the water might be too deep for them to wade. On the bank they were hindered by undergrowth, trees, and scrub grass four or five feet high. They had to surmount fallen logs, and even standing timber might be dangerous. On that very first day, the boat swung too near the bank, the topmast struck an overhanging bough, and broke square off.

Darkness fell, the exhausted crew devoured another pot of mush and tallow, rolled up in their wet blankets, and stretched out to snore.

The passenger, less fatigued, watched the deer which came out on the sandbars to escape the mosquitoes. In those days, hunters shot deer on the bars by moonlight from stands or scaffolds.

The next day they had better luck. A brisk wind caught their sail and sent them speeding up the current until midmorning. Then it died again, opposite the Tavern Rock, and the calm set them all to work again. The rocks were so-called from a cave scratched over with Indian drawings.

By the time they had tired themselves out pulling the boat through the calm, a violent windstorm with rain and thunder compelled them to lie to under the bank. There they were terrified by the violence of the wind, which threatened to throw down the trees along the bank, crush the flimsy cabin and sink the boat. When the storm abated, they made camp.

They passed Point Labadie, butting their way through the bodies of drowned buffalo drifting down, winding on a crooked course around islands, and soon encountered their first embarras. From there on at every mile or two they encountered one of these rafts of driftwood and brush matted together, sometimes twenty or thirty yards deep, which damned up the current and caused it to plunge through a spillway at one side in a raging torrent.

Then the crew threw out grappling hooks, and the towrope was wound painfully foot by foot around the capstan in the bow until the men had warped the boat up to the hook, when another hook would be fastened farther ahead and the process repeated.

It seemed as though the devil himself had everywhere planted obstacles on purpose to defeat them. Where the stream was clear of obstacles, the current was often too swift for oars, and the water too deep for poles. Where the channel was shallow, close to the bank, its course was littered with fallen trees and treacherous with snags and sawyers; but the boatmen, clinging to the cordelle with one hand and clutching the brush along the bank with the other, somehow dragged the boat forward.

So they crept on, passing one island and raft after another, fighting the current, drenched with the rain, which now began to leak through the flimsy cabin and soak the captain (known to his crew as the bourgeois, or boss) in his bunk. The crew on shore fared better, rigging up tents out of blankets.

From time to time canoes from upriver passed bringing news of their rivals' progress. And so they reached the mouth of the Gasconade over that long straight reach of about fifteen miles, where they had a fair wind and could lie on the deck while the boat sailed merrily along.

The captain was kept as busy as the men when navigation became difficult. He exerted himself like an orchestra leader or a man killing snakes, rushing from the helm to the grappling hook at the bow, or seizing a pole to help force the boat through a difficult passage, talking, singing, shouting, urging on the men like a cheer leader at a football game, keeping up their morale. He kept the men singing:

> Derriere chez nous, il y a un étang,
>   Ye, ye ment.
> Trois canards s'en vont baignans,
> Tous du long de la rivière,
> Legèrement, ma bergère,
>   Legèrement, ye ment.
>
> Trois canards s'en vont baignans,
>   Ye, ye ment.
> Le fils du roi s'en va chassant,
> Tous du long de la rivière,
> Legèrement, ma bergère,
>   Legèrement, ye ment.
>
> (Behind our house there is a pond,
>   fal lal de ra.
> There came three ducks to swim thereon:
> All along the river clear,
> Lightly, my shepherdess dear,
>   Lightly, fal de ra.
>
> There came three ducks to swim thereon,
>   fal lal de ra.
> The prince to chase them he did run
> All along the river clear,
> Lightly, my shepherdess dear,
>   Lightly, fal de ra.[1]

So they passed the Gasconade, on through the yellow waters of the Osage pouring in, to the village of Côte Sans Dessein. The village stood under this Hill without Reason—so called because it stood alone on the north bank of the river—a narrow knifelike hill, six hundred yards long. They reached the first Cedar Island, of which there were several on the river, passed the Manitou Rocks and Split Rock, reaching the narrowest part of the stream so far.

They made good time once, twenty-eight miles in one day, with the help of a favorable wind.

This good luck, the boatmen said, was due to their charity. They had found a wretched ox sunk in the quicksand under the bank. Already buzzards and crows were perched upon the neighboring trees, and wolves had been nipping at his helpless back. He was in so deep that he must have been there for days on end. But for all Lisa's hurry, they had stopped to dig him out.

Afterward their luck changed. The current was strong and the river crooked, and about where the town of Lexington is now, they found tough going. The 400-foot towline had to be tied to a tree and wound up. They had to zigzag back and forth. These crossings and recrossings were very difficult. The current, swinging from side to side, swept them down so that they lost in the middle of the stream what they had gained along the shore. At the end of a terrible day they had made just two miles!

Again a violent head wind would strike against them and keep them waiting for hours—time which Lisa philosophically spent in his bunk reading *Don Quixote*. Brackenridge, for his part, wrote in his diary, admiring the winding river with its islands, its groves and its willow borders—a scene magnificent though melancholy. They lost days in such travel around the now vanished Wizard's Island. On the 25th, they had come three hundred miles and were well beyond the settlements. They therefore rejoiced to see Fort Osage, and Lisa issued whisky for a celebration.

Fort Osage was the last place in those days that could be called a settlement. It was triangular. William Clark, the explorer, who erected it, was forever building three-cornered forts. It stood five miles above Fire Prairie, about a hundred feet above the water, at the elbow, where it had a fine view up and down the river. Here the crew and travelers were entertained and learned of a recent clash with the Osage Indians still encamped about the post. Every morning the Osages spent the hours before daylight wailing for their dead —a lament in which every man, woman, and child—and at least a thousand dogs—took part.

At Fort Osage Lisa bought some oilcloth to stretch over his leaky cabin, and laid in an extra supply of Indian goods.

Just before noon next day, the crew came aboard and pushed upriver against a head wind so strong that after two hours of backbreaking struggle, they were still within sight of the fort. They had to tie up and wait for the wind to abate. To top the evils of the day, it set in to rain. However, they all were cheered by good news. Good news: they had gained on their rival. Hunt was now only *eighteen* days ahead.

Lisa hoped to catch him six hundred miles upstream. But Hunt was pushing on at top speed, fearing that Lisa would pass him on the river and then bribe the Sioux above to ambush him.

The Sioux, of course, resented traders' carrying supplies to their enemies upriver and, having nothing themselves to trade except buffalo robes, were ready to loot or intimidate any small party coming up the stream. But Lisa's men kept on. They passed the Kaw, the Little Platte, and on May 2nd, after thirty days' journey, made camp on the Nodaway River not far from where Hunt's men had joined him to go on upriver. It was small comfort to Lisa and his crew to reflect that most of Hunt's men were fresh at this point, having waited at their ease for him to come up from St. Charles. The passenger hunted, shooting pigeons, wild ducks, turkeys, deer, bear, and snipe. The Missouri Valley in those days was a hunter's paradise.

But Lisa never rested—never allowed others to rest. There was more rain, more towing, until the worn-out boatmen began to grumble. It seemed to them they never slept, never had time to eat or smoke a pipe. Their bourgeois was without pity.

Brackenridge made a speech to cheer the men, but could not remove their discontent. On top of that, news came from two fellows in a bark canoe that Hunt had passed the Platte—and *sailed* past at that!

They replaced the broken mast and plodded on. And now the men, worked so relentlessly, began to fall sick with coughs and fevers and pains in their sides. And so, half-dead, they reached the Platte.

Here the character of the country changed. The thick woodlands and the lofty hills gave place to grassy plains with open cottonwood groves here and there along the banks, and scattered in the hollows of the uplands a few thickets. Rocks were rarely seen. The crumbling bluffs were of clay with occasional stretches of chunks of pumice, and burned-out ledges of lignite. The animal life changed also. Prairie dogs and antelope were seen. Rattlesnakes, horned toads, and magpies began to appear.

Here, according to the boatmen, the Upper River began, and they paused to hold the traditional celebration. It was all horseplay. Every man who had not come so far up the river before was put through an initiation. Unless he would stand treat to the crew, the old-timers cut his hair and shaved off his beard. Lisa wisely entered into these sports and provided the necessary cheer for his men.

Their hearts warmed by the liquor and the fun, the

men pushed on, fighting the driftwood which now swept down the stream, passing Blackbird Hill, the Omaha village, and Floyd's Bluff. They began to see frequent traces of Hunt's party. They found one camp where signs showed he must have remained several days—a cheering sight. Here they found cutoffs and multiplied bends and loops of the river, and sandbars which, in rising water, melted beneath their feet. As the water in the river was lower above the Platte, these bars were often dry, and when the wind was high, clouds of sand blinded the boatmen. These clouds of sand sweeping the bars, when seen from afar, looked like smoke. The new men readily understood why the Omaha Indians called the Missouri "Smoky River."

They now encountered dangerous rapids, and afterward high water near Vermilion Creek. Here they heard bad news from a passing employee of their own company—that *all* the tribes upriver were now hostile to the whites. In fact the man reported that he himself had been fired upon by Indians. Once more the men's hearts sank to their wet shoes.

But Lisa made oration, issued grog, started a song. The crew put sore shoulders to their poles and labored on.

With his men near mutiny and Indians preparing an ambush for him, Lisa smothered his pride and sent Hunt a letter overland begging him to wait and join forces with him. Lisa was "a very amiable and interesting young man"— with as much discretion as valor.

Hunt, having more men, was less concerned about the danger from Indians. Blandly he pocketed the letter, urged his boatmen on, and sent word to Lisa that he would wait for him at the Ponca village.

This seemed reasonable enough to Lisa's party, and wonderful news. What a relief! The weary men relaxed— but only for a moment.

Lisa took no chances. The Indians might attack at any time. He forced the tired crew to push on faster than before, still climbing that relentless river, dragging their boat now *by night* as well as by day. They hoped to slip past the hostile Indians unnoticed.

Contrary winds beset them, and Captain Lisa, standing in the bow of the boat with his grappling hook, went overboard and narrowly escaped being drowned. The boat drifted downstream. Then Brackenridge, trying to fill the place of the captain by handling a pole, fell overboard from the stern and came near being sucked under by the swift current. By good luck he grabbed the sweep and pulled himself back into the boat, little the worse for his ducking.

Lisa needed fresh meat to encourage his men. So the hunters went after antelope. An Indian heard their gunfire and appeared on the bluff. Then it was necessary to stop and hold council with the savages. But that night, while the Indians slept, Lisa's men pushed on upriver, leaving them far behind. Of course the Sioux might ride overland and overtake them, so now they dared not stop even to sleep.

When Lisa reached the Ponca village, Hunt was not there. They did not know what to make of that until two deserters from Hunt's party turned up and explained that Hunt, on learning that Lisa and his men were so near, had sent a false message to delay them. He was *not* waiting, but making frantic efforts to run away from them. Brackenridge and Lisa were inclined to think that Hunt was judging them by himself and that he would not hesitate to set the Sioux upon *them*.

Disheartened and frightened by this bad news, the boatmen could only push on harder than ever. Lisa decided to risk the hazards of sailing by night. Fortunately the moon was bright and the water middling high. Still, if he ran aground, he might lose his scalp. He sailed almost all night.

Next day, May 29th, they passed beautiful Cedar Island, and on the point saw where Hunt had camped— the fire was still burning!

Now they knew they were close upon him. The hopes of the voyageurs rose once more. The wind was favorable, and they sailed steadily on, making seventy-five miles within twenty-four hours.

At twelve o'clock they reached the Grand Detour, or Great Bend, then thirty miles around, though little more than a mile across the neck. Here Fortune favored them, for as they sailed around the Bend, the wind miraculously changed again, shifting with the course of the river so that it carried them smoothly almost the whole way round! In this way they gained a *whole day*.

Early on the morning of June 2nd, while the spent crew lay on the riverbank, too tired even to get up for breakfast, Brackenridge, taking his rifle, climbed a bluff nearby. Beyond the point, hardly a mile ahead, he saw the camp and boats of the Astorians! He ran back to tell Lisa.

Up sprang the crew, forgetful of their weariness; quickly they boarded the boat. Out went the oars, as the craft swung into the river, ignoring three hundred Ree Indians who hailed them from the bank. At eleven o'clock that morning Lisa's keelboat swept grandly round the curve into view of the Astorian camp. Even those three hundred Rees could not have made such a racket as the yelling and singing of Lisa's triumphant crew.

> There came three ducks to swim thereon,
>   fal lal de ra.
> The prince to chase them he did run
> All along the river clear,
> Lightly, my shepherdess dear,
>   Lightly, fal de ra.

They had won—and they had established a record for keelboats: nearly 1,200 miles up the Missouri in 61 days.

Brackenridge jumped ashore and grasped the hand of his friend Bradbury. These two men conspired together, combined forces, and talked Hunt and Lisa into a truce. The two parties journeyed on together.

Hunt led his Astorians over the Rockies and founded Astoria at the mouth of the Columbia. Lisa found his partner Henry, made another fortune, and remained the chief trader of the Upper Missouri until his death nine years later.

He lived to see the first of the steamboats reach the Platte—the steamboats which were to drive keelboats from the river. But his record stands. No one else ever drove a keelboat upriver such a distance at the rate of more than eighteen miles a day.

CHAPTER III

Steamboat 'Round the Bend

ANY MAN—that is, any man who was young, strong, active, alert, aggressive, determined, persistent, patient, brave, and eloquent—provided he had a good singing voice and a sufficient crew of seasoned, amphibious rivermen, could take a keelboat up the Missouri River. But if he wished to have a boat take *him* up, he bought passage on a steamboat.

Of course even a big steamboat could hardly carry more than four hundred passengers, whereas there was no limit to the number of men who might go up the river with a keelboat; most of them walked all the way anyhow. But on a steamboat there was comfort, often luxury—if not always safety. Fares varied, of course, with the accommodations and the speed of the boats, but in the 1850's—the heyday of steamboat travel on the Missouri—cabin passengers paid only from 1 to 2 cents and up per mile, luggage free, and this included a berth, abundant food, and obsequious service. Moreover, a passenger with luggage had the privilege of paying his fare on arrival at his destination, so that he was free to quit the boat at any landing where it touched.

Deck passengers paid much smaller fares, but had to feed themselves and provide their own bedding. If they had covered wagons along, they sometimes slept in the wagon beds.

From the end of the eighteenth century, the number of

people heading west up the Missouri steadily increased. Some came in the hope of making quick fortunes in the fur trade, others were settlers looking for free land, and every settler up the river meant increased freight and passenger traffic below. Moreover, every few years navigation was stimulated by some new shot in the arm: the rapid growth of trade to Santa Fe, the migration to Oregon, the various gold rushes—to California in 1849 and in the fifties to Colorado, in the sixties to Idaho and Montana, and in the seventies to the Black Hills—all brought thousands aboard Missouri steamers. Owners not infrequently paid for their boats in a season or two, and sometimes in a single trip.

As the boom in steamboating mounted, the size and luxury of the boats increased until steamboats on the Missouri could match anything floating on the Mississippi. In fact, some of the finest Mississippi steamers plied the Missouri; such was the *A. B. Chambers* on which Mark Twain was pilot in 1859. This elegant side-wheel steamer had been on the Missouri as early as 1856 when, on its first trip, it was sunk near Atchison, Kansas. Raised and repaired, it remained in service until 1860, when it was snagged and sunk, a total loss, just above the mouth of the Missouri.

The traveler, on coming down to the wharf to board such a floating palace, beheld a long, low vessel with a lofty superstructure, a flag flying at the jackstaff forward, and two tall, shiny, black smokestacks fuming into the sky. The lower, main, deck was crammed to the guards with stacks of bales, boxes, bedrolls, trunks, cordwood for the glaring furnaces, big Conestoga or Pittsburgh wagons with bright red wheels and blue beds—each stuffed with goods billed to Santa Fe and hung round with harness—besides mules, horses, hogs and sheep, chicken coops, and very likely a cage of cats—mousers for some army post or trading fort upriver. A whole battalion of scurrying roustabouts and humble deck passengers milled around the whirring windlass as freight

STEAMBOATS ON THE MISSOURI COU

...TCH ANYTHING ON THE MISSISSIPPI.

was lowered through the forehatch into the hold and shoved to its place on miniature trams running on a circular track below decks.

On the boiler, or cabin, deck above—the second story of the boat—the guards were lined with passengers. These overflowed onto the hurricane deck above, where the texas, crowned by the lofty pilothouse, afforded quarters for the officers. Such a steamboat looked rather like a fancy, gilded, white wedding cake.

Going aboard, the cabin passenger mounted the stairway to the boiler deck to locate his stateroom and see his luggage safely bestowed; then he was free to prowl about and inspect the glories of his temporary home on the voyage. There he found a long saloon stretching the length of the boat—a spacious room enriched with Brussels carpets, handsome furniture, glittering chandeliers fringed with glass prisms, the whole gaudy with gingerbread woodwork, paint, and gilt equal to that of a modern motion-picture palace. This saloon was lined with staterooms on both sides, every door decorated with a painting done in oils—usually more remarkable for vigor and sentiment than for artistic taste.

The forepart of this saloon, the Gentlemen's Cabin, was reserved for men and cuspidors. Here meals for all were served on tables set up for the purpose at mealtime. At other times the gentlemen, who were "not permitted to share the society of ladies and to be enlivened by their charming company and refining influence," spent the time playing cards, drinking, chewing tobacco, smoking, talking politics, or whittling sticks. This segregation of the sexes on steamboats amazed foreigners, but may have had its advantages in preventing duels.

The rear fourth of the saloon (farthest from the boilers and so the safest part of the boat) was the Ladies' Cabin and reserved for them, being shut off by glass doors. Only gentlemen who had friends among the ladies on board had entree there.

By all accounts ladies simply adored traveling on a steamboat. In fact, about the beginning of the Civil War the handsomest compliment one could pay one's hostess was to say, "Your house is as beautiful as a steamboat."

Sometimes fittings salvaged from a steamboat were used to adorn private mansions. One boat which sank and was swamped in sand up to the cabin deck, was used to build a fine home at Bluffton, Missouri. Later the house was torn down, but much of the woodwork was incorporated in a new building which, I believe, still stands in that town. Passengers on a steamboat traveled in luxury.

After the Civil War, it was the custom in the large towns along the river for young men to engage reservations on such fine boats as the *Montana* and the *Belle of St. Louis* for a house party starting at St. Joseph or Kansas City and invite young ladies and their chaperons. The big boat floated down the river through glorious moonlight in an atmosphere of luxury and romance. There was always music, and the Negro cooks outdid themselves for the society folks. The young ladies wore their prettiest clothes—blue silks, Paris muslins, or yellow satins. A girl thought too young for a hoop skirt wore a "tilter"—half a hoop skirt.

They danced square dances, Virginia reels, the quadrille, and the cotillion. Respectable fathers frowned upon the "round dance," or waltz, which was thought to be too much like embracing. Only blasé young men dared to waltz, and even then held the young lady by her elbows only. It was considered shocking to place the hand on a maiden's back.

Those who remember those pleasure trips can never forget the singing of the Negroes on the wharves as the boats tooted in. Many a romance flowered amid the color and luxury of those glamorous old river boats; many a young man popped the question.

Groups would join the party at towns along the way to dance all day and half the night. It is hardly surprising that ladies enjoyed traveling on a steamboat. So did the men.[1]

Having inspected his quarters, our traveler might mount to the hurricane deck, while the last bell clanged and a black boy pounded a Chinese gong and shouted "All ashore that's goin' ashore," to watch the landing stage hauled in, hear the engines begin to thump, and feel the boat slide backward into the river to head up the stream.

Nothing in the history of travel in America has equaled the delight of a trip on such a steamer, providing good food, comfortable lodging, continual change of scene, plentiful company, adventure, danger, and sheer fun. By contrast, a transatlantic crossing on a luxury liner of our day is a dull affair.

Travelers on the Missouri did not travel to and from hotels; they traveled *in* hotels—and much the best hotels in the West. Such steamboats were equipped with elaborate bars, and they set a groaning table. When one reads their old bills of fare one wonders whether, after all, the American standard of living has improved. Everything the heart could desire was on the card and all the dishes were served to everybody. A man did not take his "choice" or order "a portion." He began at the top of the bill of fare and ate his way straight down to the bottom. If he were a man of sufficient importance, he might even be served between meals with tarts or coffee or ices by the black texas-tender who looked after the ship's officers.

The passengers on board were of many types: lawyers, doctors, ministers, merchants, fur traders and Indian agents, voyageurs, military men, cardsharps, politicians, scientists, artists and writers looking for fresh scenes, explorers, army contractors, and—not infrequently—English noblemen, or even a royal prince from Europe. The main deck was crowded with a motley horde of settlers bound for California or Oregon, Negro slaves, mountain men in black, greasy buckskins and shaggy beards, and Indians with painted faces and bright blankets or buffalo robes. Besides all these were a few laconic men with watchful eyes who had deserted from the army or were fugitives from justice or injustice in the States, runaway apprentices, counterfeiters, forgers, robbers, murderers, crackpots and fanatics. The traveler on a steamboat found no lack of variety, and soon became acquainted with other passengers where all were thrown continually together.

Moreover, there was a constant change in the passenger list as people boarded or left the steamboat where it touched. On the Missouri, boats seldom traveled after sundown, and when the steamboat tied up for the night at a town or an army post, traders, army officers, and other important people came aboard to be entertained by the captain at the bar, to share in the merrymaking and dancing in the saloon, or to listen to the music provided by the boat's black orchestra or band. If these delights palled on the passenger, he could always go ashore and seek entertainment more to his taste.

By day, when the steamboat was under way, there was always something to see. Scenery slid by—towns and ferries and bridges, pioneer log cabins with a barefooted woman rocking in her chair on the stoop, Indians and buffalo, antelope and bighorns and bears, Mackinaws and broadhorns coming down the river.

On the main deck below, the antics of the polyglot deck passengers and the fights of roustabouts and deck hands provided amusement and opportunity for wagers. On the boiler deck there was usually a poker game or two in progress.

A privileged passenger might sometimes hope to go up to the pilothouse and watch the lordly pilot negotiate a difficult chute or crossing. At every landing where fuel was available, the roustabouts hustled cordwood aboard for the furnaces, which consumed eighteen or twenty cords a day. Occasionally the boat would pass the wreck of a sunken steamboat—a sight which gave passengers an exciting anticipation of possible danger ahead.

Weather, too, offered great variety on the Missouri. Thunderstorms roared and fierce hailstorms pelted the boat, to the terror or amazement of the passengers, and almost every other day terrific gales—sometimes carrying clouds of sand—blew up or down the river bed, even when the wind was not particularly strong ashore.

And the stream was winding, not to say crooked, so that there was always something new to look for around the bend. There was a continual succession of islands large and small, and every few miles a tributary coming in from one side or the other to add its water and its silt to the Missouri.

Places along the river rejoiced in picturesque names: Musick Ferry, Rattlesnake Springs, Wolf Point, Alert Bend, the Coalbanks, Peru Cutoff, Painted Woods, Cow Island, Blacksnake Hills, the Devil's Race Ground, Portage La Force, Mule's Head Landing, Standing Rock, Hole in the Wall, The Ramparts, Square Butte, Elk Horn Prairie, Brulé Bottom, Fort Mandan, The Stonewalls—the list went on forever.

The Missouri had a terminology all its own. Thus, a "point" on the Missouri meant a small copse—"a patch of woods" (to use Audubon's phrase)—or wooded promontory. On other rivers when people speak of a "crossing," they mean a place where they can cross the river. But on the Missouri a "crossing" means a place where the current crosses the bed from one shore to the other.

Such crossings were always difficult and often dangerous. The depth of the water was uncertain; but whatever the depth, the boats had to cross the current and, if it was powerful, were likely to be swept back downstream.

The flood-plain of the Missouri may be from two to six or even fifteen miles wide. Through much of its course, it is from 150 to 300 feet below the general level of the surrounding country. Thus the river has plenty of room to work and play in. Its only permanent boundaries are the bluffs. Any permanence of the lower banks is purely accidental.

Most of the accidents to boats happened at the bends of the river, and many of these bends have historical names —some of them referring to these very accidents.

Thus Cora Bend is named for the steamboat *Cora* which sank there. Tabo Bend near Lexington, Missouri, is an example of American folk etymology, having once been French: Terre Beau. Sheep Nose Bend takes its name from the shape of the bluff there, while Box Car Bend gets its monicker from the fact that scores of boxcars were once thrown into the river at that point to keep it from washing out the railroad tracks. Pelican Bend was once a resort for those grotesque birds. Near the mouth of Grand River is Bushwhacker Bend, where Confederate forces attacked a Union gunboat during the war between the states. Arrow Rock was once a slave market where visitors could see the old slave auction block which still stands on the bluff in front of the tavern. Because of this, the bend below the town is known as Nigger Bend. Great Bend, or Grand Detour, explains itself, referring to the big bend below "the Fork"—the mouth of the Cheyenne River.

The most dangerous bends, where snags accumulated, were Augusta Bend, Brickhouse Bend, Malta Bend, Osage Chute, Bonhomme Bend (in which many a magnificent steamboat was wrecked). That was a wicked river.

Rivermen on the Missouri looked with scorn upon those who had only navigated the placid Mississippi. It took a man with hair on his chest to pilot a boat up the Missouri. It wrecked nearly 450 steamboats during the years when boats were numerous on the river, though in 1858, the banner year of steamboating on the stream, there were somewhat less than a hundred boats in service.

Snags were by far the greatest danger, causing nearly half the wrecks. Fire, in those days of candles and oil lamps, accounted for a good many, with ice an equal or greater hazard. Some boats were wrecked on rocks, others against bridges. The high winds which swept up and down the river valley accounted for others; while eddies or overloading or collisions swamped a few.

In the spring, when the ice first broke up with a mighty sound of cracking and splitting, the captain would chop out his boat and break a passage by main force with his yawl out into the current. Even there, floating ice was a continual hazard as it bobbed and bounced against the hull and beat upon the paddle wheels. Frequently it was necessary to stop the engines dead to keep large blocks of ice from shattering the blades. The continual friction of the ice against the boat caused an incessant noise "like the beating of grain on a threshing floor."

On the Upper River, ice would sometimes be piled up on the shore in huge blocks, miniature icebergs, in the lee of which a man might make camp to protect himself from the bitter wind. As the river went down, it often left blocks of ice lodged in treetops along the stream, and these remained, slowly melting away, for days after.

Sometimes ice formed about a boat and held it prisoner all winter away up the river—crushed it or pushed it up on the bank, where it was left high and dry. Sometimes a boat ran hard aground on a bar when the river was falling, so that there was no chance to get it off that season. Then the ice coming down in the spring would chop it down—a total loss.

And there were frequent floods. The prankish river was always cutting up. Once or twice it moved into some town along its course, depositing a steamboat, complete with cargo, in the middle of Main Street.

It was the universal custom to load the boat more heavily forward than aft, so that the forepart sank a foot deeper into the water than the stern. If loaded too heavily aft, the boat might run aground so far that it could not be backed off.

Most steamboats were equipped with a pair of spars to be used when the boat ran aground. "When she became lodged on a bar, the spars were raised and set in the river bottom, like posts, their tops inclined somewhat toward the bow. Above the line of the deck each was rigged with a tackle-block over which a manila cable was passed, one end being fastened to the gunwale of the boat and the other end wound around the capstan. As the capstan was turned and the paddlewheel revolved, the boat was thus lifted and pushed forward. Then the spars were re-set farther ahead and the process repeated until the boat was at last literally lifted over the bar. From the grotesque resemblance to a grasshopper which the craft bore when her spars were set, and from the fact that she might be said to move forward in a series of hops, the practice came to be called 'grass-hoppering.'" [2]

When one of those wooden boats caught fire, it was likely to be a total loss. A smart captain, however, might sometimes save his cargo and his passengers by chopping a hole in the bottom and scuttling the boat so that it sank until the water rose high enough to put out the fire. If the water was not too deep, he might then plug up the hole, pump out the water, and float the boat again.

If, on the other hand, the boat was "bilged" by a snag, the captain might deliberately run it aground on a bar to prevent sinking and so save the cargo. There it might be caught by a sudden rise, swept off, and sunk. Sometimes a boat sank to the hurricane deck within sixty seconds. More usually, the water was shallower and the sinking occupied some time.

One of the main dangers in those days of iron boilers was that these might blow up. The water of the river was so impregnated with mud that this was an ever-present danger; and in the early days of steamboating, it was the custom to take off one end of each boiler every day to admit a man to clean it. Afterward a tube was adjusted to the boilers to form a passage through which the mud could be blown out. If the mud were not thoroughly cleaned out, the history of the boiler might be expressed in three words: crusted, rusted, busted! The current of the Missouri was often so swift and the channel so narrow that only a tremendous head of steam would push a boat against it. If the captain, determined to push on, dared to hang a keg of nails or a flatiron on the safety valve, the whole shebang might be blown to kingdom come.

The worst tragedy of this sort on the river was probably that near Lexington, Missouri, when Captain Francis T. Belt, impatient at repeated failures to breast the current, ordered full speed ahead. The engineer cracked on the steam and blew the steamboat *Saluda* to glory. The boat was loaded to the guards with Mormon emigrants. How many perished is uncertain, but more than one hundred bodies were recovered and buried in the cemetery there.

In addition to the dangers of navigation, there was always a chance of disaster during the night when the boat was tied up or lying to. Then the caving bank or trees might fall upon it, ice might damage it, or fire might take it. Members of the crew, sneaking into the hold with a candle to tap a whisky barrel, might get drunk and set fire to the boat; or some thief, having murdered a passenger for his gold dust or wallet, might set fire to the boat to destroy the evidence of his crime. Sometimes the cargo itself caused disaster—as when a load of broncos became frightened and stampeded to one side of the vessel, causing it to turn over and sink.

Such hazards might well cause a nervous passenger to worry, as in the classic story:

The timid old lady approached the captain of the steamboat. "Captain," she said anxiously, "they say a great many men have been drowned in this river. Is that true?"

The captain smiled reassuringly. "My dear madam," he said, "you must not believe everything you hear. I assure you I have never yet met a man who had been drowned in the Missouri River."

The captain was in command while the boat was tied up, and in addition acted the part of mine host on his floating hotel; but once his steamboat was in motion, the pilot was absolute autocrat and king of the craft.

He stood in lonely majesty in his lofty pilothouse with a clear view of the river fore and aft, turning the handsome wheel or pulling the tiller rope, or the brass knob which rang the bells as a signal to the engineer, working the boat into invisible marks, listening to the calls of the man who sounded the river with a long, painted stick, steering between snags, avoiding drifting logs, sawyers, rafts, negotiating crossings, giving a masterful exhibition of skill, judgment, and experience.

For many years there were no buoys or lights on the Missouri, and the changing stream was forever confronting the pilot with new problems. In all the 2,200 miles up to Fort Benton, the pilot could always be sure that the river would *not* be as he had seen it on his last trip. Sometimes he had to lie to because of fog, or his spars or his tiller rope might break, or he might have to lighten his cargo and then take it aboard again. He had to figure on a varying efficiency of the engines by reason of a varying quality of fuel. If he could get no ash or red mulberry or oak, and had to depend on cottonwood or lignite, the engineer might have to throw resin or tar or spoiled sowbelly into the furnaces to generate sufficient steam to get round the bend.

It is too bad that no steamboat was ever named the Red Queen, the one who ran the race with Alice when she went through the Looking Glass. Alice, you remember, seeing that they never passed anything no matter how fast they ran, observed that in her country you generally got to somewhere else if you ran fast for a long time. To which the Red Queen replied, "A slow sort of country . . . now here, you see, it takes all the running you can do to keep in the same place . . ."

Going downstream was equally dangerous, since the boat traveled three or four times faster than it could go up.

Old rivermen are fond of tales that emphasize their skill in navigating the Missouri—tales at the expense of pilots on milder streams. There is the yarn about the steamboat pilot who went blind. He could no longer take a boat up the Big Muddy. But he was not downhearted. Said he, "They say the Missouri is nothing but mud—too thick to drink, too thin to cultivate—and claim that the Missouri *pollutes* the clear Mississippi. But the fact is that the Mississippi is the sinner. The Missouri is full of honest grit until it meets the Mississippi. Then it is the Mississippi which pollutes the Missouri, turning that honest grit to stinking slime. It is a plain case of Samson meeting Delilah.

"You think I am through because I am blind. No such thing. I'll be a pilot on the Mississippi. I can *smell* my way up that stream."

And then there is the story of the rootin', tootin' Missouri pilot who saw the channel ahead blocked by one of those big Mississippi steamboats and could find no water deep enough to carry him around. He was in a hurry; he could not wait on the slow progress of that cautious Mississippi pilot. So he just got out his spars and "grasshoppered" over the bluffs for two or three miles until he had passed him!

CHAPTER IV

Long's Dragon

IT IS not surprising that pilots on the Missouri made good money. A licensed pilot could write his own ticket, for profits were great. Captain J. N. Hanson, in his excellent book *The Conquest of the Missouri*, tells how Pilot Oldham, on being asked what he would take to pilot the *Moses Greenwood*, a small side-wheel packet, from St. Louis to Weston, Missouri, gave his price as $1,000. The captain demurred, but a few days later, unable to get another pilot, hurried round to Oldham to accept his offer. The pilot blandly explained that he had a date for a picnic that afternoon, and in spite of the captain's protests, kept him waiting until the picnic was over.

Some pilots received much larger sums, since no boat could legally stir without a licensed pilot in charge. However, owners seldom begrudged this high pay, since a good pilot was absolutely essential to the safety of the property and people on board. Freight rates were high, and no insurance company would insure a boat against explosions. Some pilots drew $1,000 a month; one, mentioned by Chappell [1] made a quick trip "to the mountains" (Fort Benton) for a fee of $2,100.

Steamboats carried other things besides people and freight up the Missouri. They carried smallpox, influenza, and Asiatic cholera. Thus the packet *St. Ange,* Captain La Barge, in June, 1851, which carried the artist Rudolph Friederich Kurz and Father De Smet, was "a floating hospital without a doctor" on which men who had been well only two hours before fell into convulsions and died. The boat tied up each evening to bury the dead by torchlight. Father Van Hocken died before Kurz could finish his portrait sketch for Father De Smet.

Steamboats reached the Missouri very early. Robert Fulton had demonstrated his "floating sawmill" on the Hudson in 1807. The first steamboat reached St. Louis on the Mississippi in 1817. Then the great question was: Can steamboats navigate the Missouri? Is the river navigable?

That question has been asked at least once a day ever since, and opinion is still divided on the matter. Some shout a loud NO. Others, taking a more hopeful view, declare that it *is* being navigated now; some say, more cautiously, that it *has* been navigated; others purse their lips and shrug and opine that it *could* be navigated; while some discouraged souls merely hope that some day it *will* be.

Let George Fitch speak for the men of faith: "Of course the Missouri is navigable. The trouble is that those who have tried it have spent too much time trying to change the river to conform to the steamboats when they should have been making over the steamboats to conform to the river. The Missouri River steamboat should be shallow, lithe, deep-chested, and exceedingly strong in the stern wheel. It should be hinged in the middle and should be fitted with a suction dredge so that when it cannot climb over a sandbar it can assimilate it. The Missouri River steamboat should be able to make use of a channel, but should not have to depend upon it. A steamer that cannot, on occasion, climb a steep clay bank, go across a cornfield and corner a river that is trying to get away, has little excuse for trying to navigate the Missouri." [2]

Clearly, what our grandfathers needed was a boat on wheels, an amphibious army "duck"—with armor against snags, an airplane propeller for additional power, and machine guns for the Sioux. It is surprising that they did not work out something of the sort, seeing that the pioneers already had a boat on wheels—the prairie schooner. The bed of the schooner was a watertight boat, often used to ferry rivers; a small steam engine and a stern wheel would have sent the schooner up the current and over the bars in fine style! But the boatmen and the bullwhackers never got together on the matter. Each stuck to his own ideas, and progress halted.

Still, American inventive genius groped and struggled to solve the problem. Abraham Lincoln patented a device for lifting steamboats off submerged bars, a device most accurately described as inflatable balloons of cloth attached to the sides of the boat. But Honest Abe's water wings never caught on; so far as is known, nobody ever used them. If anybody had, the contrary Missouri would have found a way to snag or deflate them somehow.

Was the Missouri navigable?

In 1819 Colonel Elias Rector took a chance and organized a steamboat company to find out; he bought a steamboat named the *Independence* and paddled up the river.

The *Missouri Intelligencer*, May 28, 1819, thus reported the momentous event: "With no ordinary sensations of pride and pleasure, we announce the arrival, this morning, of the elegant STEAMBOAT INDEPENDENCE, Captain NELSON, in seven *sailing* days (but thirteen from the time of her departure) from St. Louis, with passengers, and a cargo of flour, whiskey, sugar, iron, castings, etc., *being the first Steam Boat that ever attempted ascending the Missouri.* She was joyfully met by the inhabitants of Franklin, and saluted by the firing of cannon, which was returned by the *Independence.*

"The grand *desideratum,* the important fact is now ascertained, *that Steam Boats can safely navigate the Missouri River.*

She had made 200 miles in eighty-four sailing hours." [3]

The *Independence* went up beyond Franklin, even to Chariton—"a proud event in the history of Missouri." Both towns celebrated. At Franklin a dinner was held at which no less than twenty-three toasts were drunk, including of course one to General Jackson. Optimists proclaimed that the time was near when a trip to the Pacific would be more commonplace than a trip to Kentucky or Ohio had been twenty years before.

Later the same year a military expedition under the command of Major Stephen H. Long set out from St. Louis to establish posts along the river, and to make scientific observations. There was great excitement throughout the valley when his boats arrived at St. Louis. A story went the rounds that the government had "ascertained" that a passage through the Rocky Mountains existed at the head of the Missouri and that only a distance of five miles separated those headwaters from the headwaters of the Columbia flowing to the Pacific. The story went that the *Western Engineer,* Long's "flagship," was to be taken to pieces, carried over the mountains, rebuilt and floated down the Columbia. Thus Americans would "traverse the continent by water." It was the old dream of a Northwest Passage.

But all legends aside, Long's "flagship" was sufficiently remarkable. The St. Louis *Enquirer* for June 25, 1819, describes it:

"The bow of the vessel exhibits the form of a huge serpent, black and scaly, rising out of the water from under the boat, his head as high as the deck, darted forward, his mouth open, vomiting smoke, and apparently carrying the boat on his back. From under the boat, at its stern, issues a stream of foaming water, dashing violently along. All the machinery is hid. Three small brass field pieces, mounted on wheel carriages, stand on the deck; the boat is ascending the rapid stream at the rate of three miles an hour. Neither wind nor human hands are seen to help her; and to the eye of ignorance, the illusion is complete that a monster of the deep carries her on his back, smoking with fatigue, and lashing the waves with violent exertion." [4]

Apparently her deck was higher above the water than was usual in river boats. She also had a mast to ship whenever necessary. She drew twenty inches of water.

Long's four boats did not all set out together, and the other three, not having been built especially for the Missouri, failed to get far. The clumsy *Thomas Jefferson* struck a snag in Osage Chute and is remembered as the first steamboat wreck on the river. The *R. M. Johnson* and the *Expedition* got no farther than a point near Atchison, Kansas, where they went into winter quarters: the first reached the mouth of the Kaw, the other stopped at the first Cow Island, some ten miles above Leavenworth. Their progress was slow. They made only 350 miles in 76 days, so that

Long's keelboats, carrying troops, had to wait for them at Fort Osage, and then continued on their own to Council Bluffs, making 270 miles in 23 days.

But the *Western Engineer,* known popularly as "Long's dragon," fought its way safely up to the new Fort Missouri near Council Bluffs and moored for the winter nine miles below Cantonment Martin.

In those days, when the principal business on the Upper River was in furs, the fur companies, each of which aspired to be a monopoly, were reluctant to help outsiders up the stream. From the beginning each handled its traffic in its own boats. In 1831, Pierre Chouteau built the steamer *Yellowstone* especially for the navigation of the Missouri. Its first cargo consisted of Indian trade goods. The boat set out from St. Louis for the mouth of Yellowstone on April 15th, but only succeeded in reaching Fort Tecumseh at the mouth of the Bad (Teton) River, then called the Little Missouri, some thirteen hundred miles up. There the boat discharged her cargo, took on a full cargo of buffalo robes, furs, and peltries—not to mention ten thousand pounds of buffalo tongues—and dropped down to St. Louis. The following year the *Yellowstone* managed to reach the mouth of the river for which it was named.

Those early river steamboats were by no means the floating palaces found on the Lower River in the great days of that traffic just before the Civil War. The *Yellowstone* was a small affair, only 130 feet long, beam 19 feet, with a hold of 6 feet. It was a side-wheeler with only one engine. It had a lower, or main, deck, a cabin, or boiler, deck, and two smokestacks; but there was no texas. When the engine failed—as frequently happened—the crew propelled the boat with poles. The fuel burned was wood, and this had to be cut by the crew and lugged on board as the boat mounted the river. Prince Maximilian von Wied, who traveled on the *Yellowstone* in 1833, was quite envious of the better accommodations on the newer steamboat *Assiniboin* (built the winter before) which had two cabins, better lighted and more agreeable than those on the *Yellowstone*. The stern cabin of the *Assiniboin* had ten berths, the fore cabin twenty-four, and the crew had a room of their own between decks. Karl Bodmer, the artist who accompanied the prince, has left us two pictures of the *Yellowstone*. It had a rough voyage.

Yet there was plenty of fun on steamboats. When the boat landed at an Indian village and the natives came cautiously down and peeped over the bank to see what befell their chiefs who had gone on board, the captain of the boat would let loose a sudden discharge of steam from the escape pipe upon them, so that they were instantly thrown neck and heels into a pile—men, women, children and dogs, old and young!

The Indians held that the "fireboat" was a living creature. They called it "big medicine canoe with eyes," saying that "it sees its way and takes the deep water in the middle of the channel." They were also impressed by the firing of the twelve-pounder and the eight-pound swivels which some of the boats carried up. The coming of the steamboat was such a portent that numbers of Sioux along the Missouri were given the personal name "Steamboat"—which their descendants still carry.

When in wild country, if the steamboat was held up for any reason or had to round a considerable bend, it was always possible for a passenger to take his gun and go ashore to find a little sport or bring in some meat for the mess. There were Indian villages to visit, great hills to climb or sketch, old-timers ready to spin yarns, and a good chance to swap trinkets, blankets, powder and lead, or whisky with the savages.

In the first years of navigation on the Missouri, whisky formed a regular part of the rations of troops stationed along the river. Indeed, judging from old requisitions of the quartermaster, the diet of the troops must have been rather like that of Sir John Falstaff—a pennyworth of bread to an intolerable deal of drink. Not infrequently one finds a requisition for "8,000 gallons of good proof whisky." The traveler up the Missouri did not lack refreshment.

Today one sees few boats on the Missouri. But the river is still a highway. Twice a year waterfowl by the million migrate up and down the stream.

All early explorers expressed their amazement at the number of waterfowl along the river. Lewis and Clark continually refer to the geese and ducks nesting along the stream. When the first settlers moved into Dakota, the wild geese kept up such a racket at night that sleep became impossible. A similar thing is happening this very week as I write, while a six-inch snow has held up the spring flight near Sioux City.

The numbers of the birds on this great flyway along the river from Mandan, North Dakota, to Kansas City and back vary with the weather, the feed available, and the persistence of the hunters. From 1933 to 1938, during the severe drouth, many of the marshes and sloughs dried up, limiting the breeding grounds of waterfowl so that the bird population dropped to an all-time low. But Canada and the United States have co-operated in shortening the hunting season, lowering the bag limit, and establishing preserves and refuges for breeding grounds. After the war began, ammunition was hard to get, and the hunters were hurrying to the far corners of the earth after bigger game. As a result, the bird population has largely increased. This year ducks and geese pass over by the million.

It is believed that, at certain times, there are more ducks on the Missouri River than on any other stream on the continent. Much of this great river flyway is closed to shooting at all seasons. In order to protect the birds en route, an armed guard of federal and state game wardens travel with the flight from Louisiana to Canada. There the Canadian police take over.

CHAPTER V

Princess of the Missouri

THE Missouri is a river of ghosts and graves. About the mouth of nearly every tributary there are mounds and vaults, earthworks and village sites containing the bones and pipes, the weapons and pottery of prehistoric rivermen. Some of these may date back to the Crusades or before, showing how long the valley has been inhabited. Of these, perhaps the most remarkable is the "Old Fort" on the bluffs a few miles from Miami, Missouri, a roughly semicircular earthwork, having ditches within and without, the embankments measuring 2,700 feet around. At the gateways, the walls overlap, forming a narrow passage in which enemies trying to enter would be under the defenders' fire at close range. The walls enclose some forty acres.

Besides these ancient village sites, there are uncounted others of more recent date, built and abandoned by tribes found on the river by the first white explorers—whether because of war or pestilence, superstitious fears, or the exhaustion of wood used for fuel, who can say? These, too, have vanished, except for their house rings and sunken caches, their ditches and crumbling ramparts; not a stick remains aboveground.

And then there are the towns built by white men which have vanished—washed into the river, or abandoned, or left high and dry by a change in the bed of the fickle stream.

It is the fashion to refer to any group of abandoned dwellings as a "ghost town." But that is a misnomer. Where buildings remain, you have a "corpse town," rather—or, at any rate, a "skeleton town." But these deserted villages on the Missouri are *genuine* ghost towns; their sites may be known, but their houses have disappeared as though they had never been.

Scores of forts and trading posts, pioneer cabins, and Indian agencies, which once stood on the points and prairies along the stream, are missing now. Some of these, like Fort Abraham Lincoln, were erected within living memory. Of many, celebrated and important in their day, not a trace remains. On no other stream in America, probably, has the destruction of old dwellings been so swift, so complete. It seems mysterious, uncanny.

But there is a simple explanation. For years after steamboats began to go up the river, there were no regular woodyards in the country of the wild Indians. When a steamboat needed fuel, which was every day, the crew went ashore to cut it. Naturally, they preferred dry, seasoned wood, and they were in a hurry to get back before some skulking Injun loosed an arrow at them. What could be more tempting than an abandoned building? And so, many a house was torn down, its logs lugged on deck, to be sawed up or chopped up while the steamboat was under way, and then tossed into the firebox piece by piece. Deserted villages on the Missouri went up in smoke.

Yet here and there—usually out of eyeshot from the stream—we find an empty house standing where the ghosts may still find shelter. Of these, one of the most famous is on the north bank near Carrollton, opposite Waverly—the Baker House. Baker was a steamboat captain who built himself this old southern mansion sometime before 1860. One night after the Civil War, three men crashed a New Year's party there. They had all served in the Union Army, and the host, who had favored the Confederacy, ordered them off the place. A duel followed in which one man was killed, his opponent mortally hurt. Since then, these two uncongenial ghosts remain to dispute possession of the premises. Nobody else will live there.

The haunted house is not far from the mouth of Grand River.

Perhaps the first to explore and chart the Missouri above Grand River was a young Frenchman, Etienne Veniard de Bourgmond. Certainly he followed the river up to the Kaw and probably as far as the Platte. He left St. Louis March 29, 1714.[1]

Like many another Frenchman, he quickly made friends with the native people, and particularly with the Missouri tribe and the Osages, whom he describes as "a splendid race, and more alert than any other nation." He rubbed his eyes at the quantity of game and the beauty of the country. Says he: "The meadows are rolling like the sea and abound with wild animals . . . in such quantities as to surpass the imagination. All the tribes hunt with the arrow; they have splendid horses and are good riders."

In 1719 he sailed back to France, was given a royal commission as "Commandant de la Rivière du Missouri" and promised a handsome reward and noble rank on condition that he establish a fort on the river within two years' time and conclude treaties of peace with the Kansas and Padouca Indians or other enemy tribes ranging toward Mexico. The young man was afire with ambition.

Accordingly, in 1722 he reached New Orleans—only to find that no supplies or men were available for his mission.

Early in 1723, however, he went up the Mississippi with such men and supplies as he could gather. He had a tough time until he met his old friends, the Missouri Indians. That autumn he reached their village on the river with a force of forty men, and set about building his fort. No remains of the fort have been identified. De Villiers concludes that it stood on the north shore of the river about opposite the present town of Waverly in Carroll County, Missouri. Lewis and Clark thought it was on an island.

Wherever it may have been, de Bourgmond had a devil of a time getting it built. In those days the French monarch conducted his affairs according to a charming system: he sent one man out to do something—and sent two others along to keep him from doing it. The two envious commissaries with the commandant showed slight interest in the purpose of his expedition. They insisted upon first putting his forces to work building houses for themselves, so that the commandant had only three men and his personal servant for work on the fort. These commissaries would not even let him sleep in one of their houses. He had to live in a hut made of sticks with a grass roof "just as God grew these in the wood."

Yet, in spite of all difficulties, he completed his warehouse and personal quarters, a makeshift church and shelter for the priest, barracks, an armorer's cabin, and a hole in the ground for an icehouse. He built no fortifications, since his small force could not have manned them had they been built. Instead he made friends with the Missouri Indians.

The commissaries envied his influence over the natives. But their spiteful conduct only forced de Bourgmond to turn all the more to the redskins—thereby increasing their loyalty to him. Thus he made himself dominant.

Finally he threatened to return to France and complain, and so got the men and supplies he required. In June, 1724, he set out with a convoy of canoes and reached a point across the river from St. Joseph. That summer he led nineteen Frenchmen, and hundreds of warriors with their women and children, heading for the Padouca country. Their baggage was carried on traveaux drawn by dogs. He made a treaty of peace—one of the great aims of his expedition.

Having thus accomplished his mission, the commandant

prepared to return to France. He invited the Missouri, the Osage, the Oto, and the Illinois Indians to send a delegation of chiefs to the court of his Catholic Majesty, King Louis XV. It was a long journey, full of unknown perils, and wild Indians might well hesitate to accept the invitation—particularly as an Oto chief had died on a trip to visit Bienville some years before. On the other hand, they were anxious to establish a firm trading alliance with the French, having no liking for the English or the Spaniards in their country. On his first visit to the Missouri, de Bourgmond had acquired an Indian son; this boy, sent to France, had returned recently in safety. No doubt the Indians were encouraged by that circumstance, and of course it was difficult for any leading warrior to refuse to undertake such a *dangerous* enterprise. The prospect of rich gifts must also have had weight with the council.

They wanted guns, lead, and powder in order to withstand their well-armed enemies pressing upon them from the east.

Indians fear the dead and take their dreams seriously. In bidding farewell to the commandant and the delegation, their spokesman said, "We will cry for you a little so as not to dream about you if you happen to die on the way"—a statement which throws an interesting light upon the vociferous and long-continued mourning of Indians for their dead.

In addition to the chiefs and his sergeant, de Bourgmond took along the daughter of the head chief of the Missouri tribe, said to have been the mother of his son. But at New Orleans, as usual, the commandant found the officials reluctant to fall in with his plans. They turned back most of the delegation. De Bourgmond arrived in Paris with only "la sauvagesse" and one chief from each tribe.

The red "ambassadors" were taken under the protection of the Duc de Bourbon and the Duchesse d'Orléans.

The Indians, and particularly the Illinois chief, Chicagou, made a big hit at court. The king granted them an audience at Fontainebleau, and did not forget to fulfill his promise to ennoble de Bourgmond, whose new coat of arms showed on a blue field a silver mountain and a naked savage. It was unfortunate for that dynasty that there was no silver mountain on the Missouri River. Had there been one, perhaps France would be a monarchy today.

While in Paris the Indians were entertained everywhere, danced native dances at the Opera and the Théâtre Italien, and killed a deer in the Bois de Boulogne.

Chicagou, accustomed to the use of sweet grass and other delicate perfumes used by Indians, found the strong French perfumes unpleasant. To him, the courtiers "smelled like alligators." *Le Mercure de France* and other records of the time give detailed accounts of this visit. De Bourgmond was well content.

The four chiefs, in laced coats and cocked hats, accompanied the "Princess of the Missouri" to Notre Dame de Paris, where she was baptized and immediately afterward married to Sergeant Dubois, the commandant's right-hand man. Among the wedding gifts was a beautiful repeating watch, a "big medicine" which simply terrified the bride. All the Indians were loaded with gifts; Chicagou received a fine snuffbox!

Back on the Missouri River, Chicagou attempted to enlighten his people, describing to them the astonishing habits of the French, who piled one cabin on top of another as high as a big cottonwood tree, who were as many as the leaves on the tree, and who traveled in wigwams of leather.

His people thought the chief's mind had been unseated by his journey; they would not or could not believe his marvelous tales.

Finding that he was rated a liar, Chicagou determined

to live up to expectations. He told how French surgeons could replace arms and legs which had been lost or mislaid by their patients. His story apparently lingered long in the memories of the Indians of the Missouri Valley, and in time spread all over the Plains. It is said that Brigham Young was once asked by a one-armed Indian to give him a new arm, and that Brigham had a hard time convincing his own followers that it would be unwise to grant the request.[2] The Indian, it is said, insisted; he was perfectly willing to enjoy the distinction of wearing three arms in the hereafter.

Fort d'Orleans was abandoned in October, 1727. The story goes that the garrison was massacred by the Missouri Indians. However that may be, the French continued to explore the river.

For the French and the Indians naturally took to each other. Both had much in common, for both were easygoing, with plenty of dash and courage, and a great love of prestige. In Indian country, the Anglo-American tried to make himself monarch of all he surveyed; the Frenchman simply made himself at home there.

CHAPTER VI

The Big Muddy

FROM the time of the first French settlement on the Mississippi in the last years of the seventeenth century until near the end of the eighteenth, the Mississippi was, for all practical purposes, a French river. All that country about the mouth of the Missouri afterwards known as Upper Louisiana was then called "the Illinois."

Settlers were few. But by 1743 the French inhabitants of "the Illinois" were sending boats down the Mississippi loaded with their products. Already in 1745 the number of Negro slaves in the district was nearly one-third of the scanty population.

These French settlers about the mouth of the Missouri were remote from the political intrigues of Quebec, New Orleans, and Paris. They lived in tiny compact villages, safe from Indians, and without social distinction or extremes of wealth and poverty. Most of them were related by blood or marriage, and so they met upon a footing of equality. They had no aspirations whatever toward progress, education, and political freedom, but were quite content to perform the duties and enjoy the amenities of the old French régime.

They continued to live in this way in spite of the fact that Louisiana changed hands with astounding frequency.

Thus, in 1762, France ceded Louisiana by a secret treaty to Spain, apparently in the hope that this subterfuge would prevent the English from taking possession at the end of the French and Indian War. But the Treaty of Paris, signed the following year, ceded a great portion of Louisiana east of the Mississippi to Great Britain, at the same time publicly making over to Spain all of the province west of the Mississippi as well as the Isle of Orleans.

But it was not until 1769, six years later, that the Spanish Governor, Don Alessandro O'Reilly, came to take possession at New Orleans. Even then French customs and laws continued undisturbed.

After the Spaniards took over, they soon became alarmed for fear the English might invade the Upper Louisiana colony from Canada. An obvious protection against this danger was a sudden increase in population—preferably of people who were hostile toward the English.

The Americans of those days filled that bill precisely, and so the Spaniards decreed that they would grant free land to all settlers who would assume the slight expense of the survey and clerical labor involved.

This meant that any American who crossed the Mississippi could have a rich bottom-land 800-acre farm for around $40 in a country where wild meat could be had for the shooting. This was an even better bargain than that which had brought settlers to Kentucky and Tennessee. Thus the Spaniards were the first to offer free homesteads in the West. In 1783 the east bank of the Mississippi as far south as the 31st parallel passed into the control of the United States, thus bringing Americans much closer to the Spanish province.

Yet Governor Miro, though ready to wink at the Americans' nonconformity in religious matters, declined to grant them the right of self-government, fearing that they might soon declare themselves independent—as actually happened later in Texas. Humorously he declared that, on such conditions, "he would undertake to depopulate most of the

United States and draw all their citizens to Louisiana, including the Congress itself."

Americans swarmed in and filed claims on nearly two million acres. By the secret treaty of San Ildefonso, Spain ceded Louisiana back to France in 1800; but while the French delayed to take possession, the Spanish Intendant issued a proclamation forbidding citizens of the United States any further use of New Orleans "as a place of deposit for merchandise," and "free transit for ships down the river to the sea." The Congress of the United States promptly appropriated two million dollars for the purchase of New Orleans, and in 1803, President Jefferson nominated Monroe as Minister Extraordinary to aid the American ambassador Livingston in negotiating with Napoleon Bonaparte, then First Consul of France, for a treaty with the purpose of enlarging and securing American rights and interests on the Mississippi River and in the territories to the east of it. Napoleon, probably fearing that he could not hold Louisiana against the English and being in need of money, offered to sell the whole of Louisiana. At that time Americans already outnumbered French and Spanish settlers there. The Americans were moving in.

This migration was encouraged by the Ordinance of 1787, wherein Congress prohibited slavery north of the Ohio River. Slaveowners now turned to the Missouri seeking new homes. There had always been slaves there.

Among the first American comers in the last decade of the eighteenth century were Daniel Boone, his wife, Rebecca, and their two sons, Nathan and Daniel Morgan—not counting other members of the clan. St. Charles was the seat of government.

When, in 1804, Lewis and Clark went up the river, they described the town as having one principal street a mile long, parallel with the river, on which stood "100 small wooden houses besides a chapel." The inhabitants, then about

WILLIAM CLARK

MERIWETHER LEWIS

450 in number, showed the explorers "the amiable hospitality of the best times of France."

At first called Les Petites Côtes, it stood on the left bank on the first high ground above the mouth of the river. With its houses strung along a bench well above high-water mark, and its stone tower or fort on the bluff above, it was a miniature Quebec.

Thus, St. Charles is the oldest permanent white settlement on the Missouri. But in the old days there were plenty of river towns older than St. Charles—Indian towns.

Somewhere in this region once lived the tribe from which the river takes its name. People have never stopped debating what that name means.[1]

Père Marquette first put their name on the map as "8emess8rit," which has been transliterated as ouemessourit, wemessouret, and oumissouri. Nicolas La Salle, who in 1682 accompanied Robert La Salle to the mouth of the river, calls it "the river of the Missouris." Apparently Joutel, in 1687, was the first to spell the name Missouri as we do. But these variations are not the only forms of this tribal name; *The Handbook of American Indians North of Mexico* gives no less than twenty-eight spellings for it.

The Missouri Indians belonged to the Chiwere group of

the Siouan stock, but it is by no means certain that "Missouri" is a Siouan word. The tribe calls itself Niutachi. This may be Englished as "to be drowned at" and probably is a nickname referring to some disaster on the river. Among the Sioux, bands which split off and set up as tribes were often nicknamed thus; sometimes they even accepted the nickname and used it themselves. Thus the Oto ("Lechers") were so called because they split off and set up as an independent tribe after a quarrel between two chiefs over a woman.

There are those who maintain that the word "Missouri" comes from an Illinois dialect of the Algonquian stock, probably Sac or Fox; they say that it means "wooden canoe." If that is its meaning, it can hardly be Siouan. The Sioux word for wooden canoe is "can-wata," the elements standing in the natural English order.

Others suppose the name Missouri means Big Muddy—and it is a fact that most of the tribes along the stream called it by some term referring to the soil in it. I suspect that the word "Missouri" may be a corruption or dialectal form of the name given to the stream by the Sioux. Siouan languages differ a good deal.

The Sioux called the river "Mini Sose" (pronounced Min'-ny So'-say)—which means "water roiled" or "muddy"; in English order—Muddy Water. The Sioux for "muddy, big" would be "Sose Tanka"; in English, Big Muddy.

Tonty met some Missouri Indians in 1682 on the Mississippi below the mouth of the Illinois. In 1804, Lewis and Clark found them in towns south of the Platte River. By that time smallpox and war had reduced them to three hundred souls.

Some may be disappointed that this mighty stream should take its name from a tribe already almost extinct when white men first explored the stream. It might seem more appropriate to call the river the Big Siouan River. At almost every point along the river, the first explorers found Siouan people in possession.

It is true that, just above the mouth on the left bank, they found some Indians of Algonquian stock, chiefly Sac and Fox. It is true that a small tribe of Caddoan stock, the Arikara or Rees, had a few fortified towns in North Dakota in the vicinity of Grand River. Also, on the headwaters of the river, the Algonquian Blackfeet and certain tribes (Flatheads and Snakes) of Salishan and Shoshonean stock touched the stream. At one time Cheyennes and Pawnees lived on the river. But these intrusions were as nothing against the over-all domination of the river by the Siouan tribes.

Such a concourse of peoples, all of one stock, in possession of the stream from mouth to headwaters demands a Siouan name, and fortunately the name Missouri, whatever its origin, is that of a Siouan tribe. The word is colorful, sounding, and somehow suggests the rush and volume, the wildness and mystery of the stream.

The king of France in 1712 refers to the Missouri as "The River St. Philip, heretofore called Missouri." Fortunately folks on the river simply ignored the royal geographer. Most people will agree that Missouri is a more satisfactory name.

But the days of French and Spanish domination of the Missouri were soon to end. Napoleon I, fearing that he might lose Louisiana to Britain, willingly listened to the offer of hard cash made by President Thomas Jefferson, who before the purchase was completed, had already secured an appropriation of $2,500 from Congress for the exploration of the region. Captain Meriwether Lewis, private secretary to the president, was given command of the expedition. He asked to have William Clark share the command with him. Clark was commissioned a second lieutenant, subordinate to Lewis. But, disregarding military rank, Lewis made him coequal in command of the expedition. Both men are referred to always as "the captains."

They left the mouth of the Missouri on May 14, 1804, and returned to St. Louis on September 23, 1806. All along the river to the headwaters, we shall follow their trail. But nowadays we shall find much to interest us that did not exist in their time. The Missouri is not a river on which time stands still.

CHAPTER VII

Beaver Tails and Wild Honey

THE Lower Missouri has always been a land flowing with milk and honey, and the people who settled there made the most of it. They found the soil adapted to the cultivation of fruit, grapes, and berries; the country teeming with game, big and little; forests full of wildfowl and rivers swarming with fish.

The soil was deep and rich; the climate mild and warm with a long growing season, so that the river was usually open all the year as high up as Boonville. Those sandy, gravelly slopes under the sheltering bluffs along the stream were well drained, warmed by the sun, and admirably suited to the culture of the vine.

Missouri was fortunate, too, in the character of the people who settled it. First came the French with their wonderful wines and soups and salads; then the Virginians with their tradition of good living, their sugar-cured hams and golden fried chicken; the hunters from Kentucky and Tennessee led by Daniel Boone, connoisseurs of wild game and corn whisky; after them the German colonists with their wine and beer and hearty diet; and finally the middle western farmer with his beeves and hogs and turkeys, his vegetables and grain. Even before these, the Indian had fared sumptuously, and the tradition of good living has remained unbroken to the present day.

Some of the favorite dishes of our fathers in this region are no longer available: wild pigeon pie, beaver tails, wild honey.

The Italian honey bee, imported from Europe, advanced with the frontier, so that its advent in the region was a signal of warning to Indians, who knew that white men would soon follow. Bee trees, filled sometimes with barrels of honey, were found everywhere in the forests along the river. Sometimes bees swarmed on a passing steamboat. By all accounts, the wild honey made by bees along the Missouri was infinitely superior to the honey we buy nowadays. All early travelers go into ecstasies over it. The Hon. Charles Augustus Murray says "the flavor was delicious, and I ate it in quantities which would have nauseated me had it been made from garden plants, instead of being collected from the sweet wild flowers of the prairie. Our life was most luxurious in respect of bed and board, for we had plenty of provisions, besides the pheasants, grouse, etc., that we shot."

There was a lady on the frontier who, being given some wild honey, was so delighted that she inquired where it came from. She was told that it was produced by a certain insect. Being eager to produce her own honey, she wrote to friends in the East asking them to send her "a pair of bees."

Henry R. Schoolcraft penned the classic description of the greatest dainty known to the Missouri hunter:

"We were invited at supper, as a particular mark of respect, to partake of a roasted beaver's tail. . . . Having heard much said among hunters concerning the peculiar flavour and delicious richness of this dish, I was highly gratified in having an opportunity of judging for myself, and accepted with avidity the offer of our host. The tail of this animal, unlike every other part of it, and of every other animal of the numerous tribe of quadrupeds, is covered with a thick scaly skin, resembling in texture certain fish, and in shape analogous to a paper-folder, or the bow of a lady's corset, tapering a little toward the end, and pyramidal on the lateral edges. It is cooked by roasting before the fire, when the skin peals off, and it is eaten simply with salt. It has a mellow, luscious taste, melting in the mouth somewhat like marrow, and being in taste something intermediate between that and a boiled perch. To this compound flavour of fish and marrow it has, in the way in which hunters eat it, a slight disagreeable smell of oil. Could this be removed by some culinary process, it would undoubtedly be received on the table of the epicure with great eclat." [1]

To top all this, the region produced excellent tobacco, and the Germans who settled Washington on the south bank, soon set up factories to make a superior kind of corncob pipe now famous throughout America as the "Missouri meerschaum." They also manufacture zithers, so that a man can have music while he smokes. One could build a wall around Missouri and live inside it like a prince on the native products.

Washington Irving, who loved to take his ease at his inn, struck the note of this country once and for all in 1832, when he described the small French hostelry across the river from St. Charles: "With its odd diminutive bowling green, skittle-ground, garden-plots, and arbours to booze in, it reminded us more of the Old World than anything we had seen for many weeks."

All up and down the Lower Missouri there are fine old stone mansions which testify to the prosperity and ample living of old times on that stream.

Missouri even provided salt in plentiful supply at a time when salt was still an expensive luxury. Daniel Boone and his stalwart sons were among the first to develop these resources at Boon's Lick, which gave its name to all the region around about—the Boon's Lick country. They brought large iron kettles upriver, boiled out the salt around Boon's Lick Spring, and sent off keelboats full of salt regularly every few weeks.

Though a poor man, Boone was a very important one in the early history of Missouri, serving for four years (1800-1804) as syndic of the Femme Osage section of the St. Charles district of Upper Louisiana.

Daniel Boone was no lawyer, but he was an honest man. His court followed no rules of evidence and paid little attention to legal precedents. He aimed to judge every case by his own firm standards of right and wrong, and no appeal ever was made from his decisions.

Daniel, in the words of the tribute paid him by Congress, was "the man who opened the way to millions of his fellowmen." Though serious-minded, he was always pleasant and, having a clear conscience, sang and whistled a good deal. He had left Kentucky because it was "too crowded"; because he "wanted elbow room."

His creed or code is preserved to us in a letter addressed to his sister in that orthography peculiar to himself: "And what chance we shall have in the next we know Not for my part I am as ignorant as a Child all the Relegan I have to Love and fear god beleve in Jeses Christ Don all the good to my Nighbour and myself that I can Do as Little harm as I Can help and trust on gods marcy for the Rest and I Beleve god neve made a man of my presipel to be Lost. . . ." [2]

Boone made the Boon's Lick Trail, over which swarmed the backwoodsmen from Kentucky, Virginia, Tennessee—all devout believers in Andrew Jackson and democracy. These men, with their long rifles and their hunting knives, their jugs and their Bibles, knew little of refinements and lived a life more than half Indian. They could not afford to waste powder and lead, and when they drew a bead on a critter, they felt in their bones that its time had come, whether the critter was a b'ar or a cussed Injun. When they went to church with their families, they usually carried their shoes until they reached the door of the church-house to save shoe leather. But once inside, they made the building ring with their hymns and shouting.

At first they could not afford much in the way of foffuraw (as they called all civilized luxuries), and the taverns in their country were modest in proportion. In the early days of Montgomery County, Missouri, William G. Rice was elected county assessor. At the time, the county was heavily in debt. But Rice was a methodical man with ideas of his own. It is said that on most of his journeys over the county, he rode an ox, making all of three miles an hour. But however slow Rice was, he got there just the same. When his term of office ended, he had paid off the public debt and left money in the treasury.

Afterward he ran a tavern on the Boon's Lick Trail. This was his bill of fare:

Corn bread and common fixin's	25 cents
Wheat bread and chicken fixin's	37½ cents
Both kinds of fixin's	62½ cents

But the backwoodsmen did a man's work on such food, and were apt to scorn the refinements of people of a more gracious background. One of these, a colonel, staying in a pioneer cabin, on Sunday morning turned up for breakfast in a fine black suit. The whole family crowded around, never having seen a long-tailed coat before. One of the girls stroked the garment with a venturous finger, then jumped back exclaiming, "Oh! ain't he *nice!*"

The father of the family did not like so much style. "Nice, hell!" he growled. "He looks like a blacksnake that has just shed its skin."

Yet when it came to a fight, the backwoodsmen were hard to match.

The Boon's Lick country was defended by no less than seven forts during the War of 1812, when the British turned the Indians against the settlers on the Missouri. One of these, Cooper's Fort, stood not far from the river. Colonel Cooper, after whom the fort was named, was in command when he received a letter from Governor Howard warning that the Indians would attack and advising the settlers to pull out and gather at St. Louis where they would be safe.

Colonel Cooper sat down and wrote a reply which may stand on the record as a true expression of the spirit of his men:

"We have maid our Hoams here . . . & all we hav is here & it wud ruen us to Leave now. We be all good Americans, not a Tory or one of his Pups among us, & we hav 2 hundred Men and Boys that will Fight to the last and we have 100 Women & Girls whut will tak there places wh. makes a good force. So we can Defend this Settlement wh. with Gods help we will do. So if we had a few barls of Powder and 2 hundred Lead is all we ask."

Sure enough, the Indians, all afoot, attacked in force. Things looked bad. The men in the fort voted to send for reinforcements.

The nearest stockade was Fort Hempstead, some six or seven miles away. Cooper felt that he could not spare any men for such an errand. Thereupon his daughter Milly volunteered. There was only a slim chance that she could get through; but admiring her spunk, he helped her astride his horse. Said he, "Is there anything you want?"

She replied, "Only a spur, father."

He buckled the spur on her foot, drew the bolts, opened the gate, and away she went through the woods.

She came so suddenly upon the Indians that they were taken by surprise. The men in the fort heard their whoops and the banging of their guns, and gave Milly up for lost. The slow hours dragged on. Powder and hope were running low in the stockade when Cooper and his men heard the crack of long rifles and a yelling that could come only from the throats of white men. The Indians, caught between two fires, quickly retired. Milly, unharmed, came riding into the fort at the head of her reinforcements.[3]

When the War of 1812 ended, the settlers found themselves in a pretty fix. The Treaty of Ghent expressly forbade any military activity "pending treaty-making with the Indian allies of the English." This ruling left the Indians free to attack while the settlers were required to keep the peace. In fact, during the six months after the treaty was signed, half a dozen battles were fought along this part of the river.

The most desperate was that in defense of the tiny village of Côte Sans Dessein. The blockhouse, suddenly attacked, contained only five defenders—three men and two women, including Baptiste Louis Roi and his wife. The Indians, determined to wipe them out, attacked repeatedly. One of the men devoted himself to prayer, leaving the other two to do all the fighting. The women cut patches, molded balls, and loaded the few weapons in the blockhouse. It was nip and tuck for hours, with only two men to defend four sides of the building from that horde of savages. The Indians charged again and again.

"Finding they could not carry the fort by storm or siege, they resorted to the use of fire. Fastening combustible materials to their arrows, they were ignited and then shot into the roof of the blockhouse; but as often as this was done the women extinguished the fire by a judicious use of the little water they had within the building. The blockhouse stood near the river bank, but the garrison was too weak to risk a single life by going after the precious liquid, and they watched with appalling interest the rapid decrease of their scanty stock. Each new blaze was heralded with demoniac yells from the assailants; and at last the water was exhausted—the last drop in the last bucket had been used! The next instant the roof over their heads was in a blaze, and despair stamped itself upon the features of the devoted little band.

"But at this critical moment one of the women produced a gallon of milk, and the flames were again extinguished. Soon another shower of blazing arrows fell upon the roof, and it was soon on fire again. Roi and his brave comrade looked silently at each other, and then glanced sorrowfully toward their wives. They felt that their time had come, and well they knew the fate worse than death that awaited the loved ones should they fall into the hands of the infuriated savages.

"For a moment Mrs. Roi disappeared in an adjoining room, and when she came out again, her face was lighted up with a smile of triumph. In her hands she held a vessel, familiar in all bed-chambers, that contained a fluid more valuable now than gold. Again the fire was extinguished, and then the little garrison sent forth a shout of exultation and defiance.

"Three times more the roof was set on fire, but each time the mysterious vessel supplied the needed liquid, and the flames were extinguished. At last, the Indians finding themselves baffled at every turn, screamed a bitter howl of rage and resentment, and withdrew." [4]

But the settlers from Virginia and Kentucky were not only good fighters and democrats, they also believed firmly in education. It is significant that the Boon's Lick Trail runs through Columbia, the seat of Missouri's State University. Newspapers also came early to the Lower River, such as the *Missouri Intelligencer*, published in Franklin.

Franklin was founded in 1816, about five years before William Becknell opened the Mexican Trace, afterward known as the Santa Fe Trail, across the plains from that point. Very soon Franklin was second in population only

KIT CARSON

to St. Louis (population about 2,000) with a land office and a business boom. The newspaper boasted that "the public square contains two acres" and that the principal streets were "eighty-two and one-half feet wide." The editor asserted, not without some show of reason, that the town afforded "an agreeable and polished society."

Franklin was the jumping-off place for the trade to Santa Fe; and local boosters, as the furs and mules came piling in from the West, the traders and settlers from the East, looked forward to a rosy future—all but the bound boy, Kit Carson.

But the Missouri River has a habit of taking a hand in human affairs and, only twelve years after the town was founded, went on the rampage and washed it entirely away. Only the graveyard on the hill was left. Afterward New Franklin was built upon the bluffs above.

Yet Old Franklin has never been forgotten. The reason is curious, perhaps unique, in the annals of vanished cities. For Franklin's claim upon our memory lies in a single brief and futile advertisement which once appeared in the local paper there:

NOTICE: To whom it may concern: That Christopher Carson, a boy about sixteen years old, small of his age, but thickset, light hair, ran away from the subscriber, living in Franklin, Howard County, Mo., to whom he had been bound to learn the saddler's trade, on or about the first day of September last. He is supposed to have made his way to the upper part of the State. All persons are notified not to harbor, support, or subsist said boy under penalty of the law. One cent reward will be given to any person who will bring back the said boy.

(Signed) David Workman

Franklin, Oct. 6, 1826.

Boundary

CHAPTER VIII

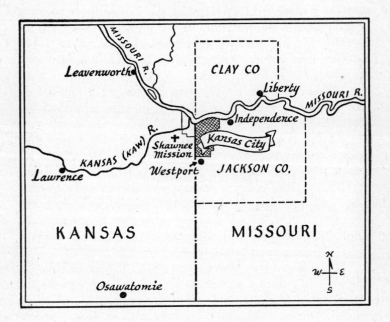

The Mormon Migration

On its upper reaches in the state of Missouri the river was the scene of many conflicts—not merely with Indians but between whites. The river below Franklin was said to flow with milk and honey. That above might have been more aptly characterized by blood and thunder. The earliest civil war within the state was that between the Mormons and the settlers who had preceded them.

In 1830 Joseph Smith had published the *Book of Mor-*

mon and organized, in New York State, the Church of Jesus Christ of Latter Day Saints. The *Book of Mormon* contains a number of promises to the Lamanites, who were identified with the American Indians and were believed to be descendants of Israelites supposed to have come to North America long before. Accordingly, the Mormons soon sent out a mission to the Indians.

Having stopped by the way to convert a number of people in Ohio, the mission to the Lamanites arrived at Independence, in Jackson County, Missouri, early in the year 1831. Here two of the men set up a tailoring business to finance the group, while the other three went upriver to preach their doctrine to the Shawnee and Delaware Indians.

The Indians, entirely tolerant and always ready to consider any promise of supernatural power, welcomed the strangers. But the Indian agent was not so complaisant. Since these Mormons had no authority to enter the reservation, he promptly expelled them.

Soon after, the Mormon prophet, Joseph Smith, announced a revelation by which he and Sidney Rigdon, another leader of the new faith, were commanded to go to western Missouri and consecrate "the land of Zion." Reaching St. Louis, they walked up the river and arrived at Independence in July.

Smith, like everyone else, was delighted with the country. The Mormons promptly bought land in and around Independence and held appropriate ceremonies preliminary to the erection of their proposed temple.

By this time the steamboats had learned their way farther and farther up the stream, and Independence had become the jumping-off place for the overland trail to Santa Fe. It was entering upon a ten-year boom and was filling up with energetic people, largely drawn from Tennessee and Kentucky.

Now the Mormons have always had a liking for doing

THE MORMON MIGRATION

things according to plan, and they soon made blueprints for a community of 20,000 people. By the middle of 1833 they had established several "stakes" in Jackson County, and already were one-third of the whole population.

From the beginning the Mormons were not popular with the people on the south bank. Theocracy had never gained a foothold in the South, and the Mormons were unsympathetic to the original settlers in a number of other respects: Mormons abhorred slavery, they made friends with the Indians, they were clannish, aggressive, exclusive, and seemed to feel that they were holier than their neighbors. Their zeal and thrift and their outspoken contempt for the institutions of all others—whom they referred to as "Gentiles"—were hard for Missourians to stomach. Moreover, the Mormons were a prosperous group, growing by leaps and bounds, and attracting to their ranks many "furriners" from overseas. They made no secret of their intention to build a wholly Mormon community and to occupy all that part of the country, where they planned to build "the most magnificent temple on the face of the earth." And when it was realized that they would not touch tobacco, liquor, or even tea or coffee, it seemed to the other settlers that they were hardly human.

They had already had some difficulty in Ohio, where, it was said, Joseph Smith had been tarred and feathered.

At Independence they established a store for the faithful and cultivated some two hundred farms, making extensive improvements. They also published a paper in which uncomplimentary remarks were made about the "ungodly" Gentiles.

At that time polygamy had not been openly advocated by the Mormons, or hostilities might have begun much sooner than they did.

In July, 1833, an angry mass meeting was called in the courthouse at Independence, which drew up an address declaring that no Mormon should in future settle in Jackson County and that those already there must sell their property and move out. The meeting required the editor of the Mormon paper, *Morning and Evening Star,* to shut up shop forthwith. They called upon the Mormon leaders to forbid future immigration to the county, and finally warned the Mormons that, should they fail to comply with these requirements, they had better "refer to those of their brethren who have the powers of divination or of unknown tongues to inform them of the lot that awaits them."

Having adopted this address, the meeting recessed for two hours while their committee waited upon the Mormons. Not receiving immediate acceptance of their terms, they passed a resolution to demolish the Mormon printing office. Immediately after, a mob destroyed the building in which the paper had been printed and threw the type into the river. They tarred and feathered two or three Mormons, including the presiding bishop, gave one or two others a cruel lashing, and burned several houses.

Three days later the Missourians, all boots and whiskers, heavily armed and carrying a red flag, *persuaded* the Mormon leaders to agree to induce their followers to move out of the county before the end of the year.

The Prophet was then in Ohio, and his harassed followers sent one of their number to consult him. Naturally the Mormons were reluctant to submit to such highhanded treatment. Hoping that the law might protect them, they appealed to the governor of Missouri. Governor Dunklin looked upon their petition with favor, and the Mormons employed a staff of prominent lawyers, among whom was Alexander W. Doniphan, afterward a famous leader in the Mexican War. Soon after, the Mormons publicly proclaimed that they intended to remain in Missouri.

There was another Mormon settlement on the Big Blue River ten miles west of Independence. Here in October a

THE MORMON MIGRATION 89

mob destroyed twelve houses, frightened the women and children out into the night, and lashed a number of men. On the night after, the Mormon store in Independence was looted. The local justice of the peace refused to take any action, and when the Mormons caught one of the mob and haled him before the justice, he jailed them for false arrest.

The Mormons, finding that the courts would not protect them, took arms and patrolled their settlements. Nevertheless, early in November a number of Mormon settlements were attacked, and, during the trial of the Mormons arrested in Independence, the prisoners might have been lynched but for the courage and swift action of the sheriff.

By this time the Mormons saw that it was no use. For two days they crowded the ferries, taking refuge on the north bank of the river with little more than the clothes they wore, huddling around their fires, trying to find lost members of their families, and sharing what little food they had. The Missouri, always taking a hand in whatever was going on, brought on a furious rainstorm which drenched these poor wretches to the skin. By the end of 1833 not a Mormon remained in Jackson County.

The kindly people in Clay County across the river befriended the refugees, gave them employment and shelter. For three years there was peace on the Missouri. Meanwhile the Mormons, taking heart, petitioned the governor to restore their lands and protect them. They also requested that they might organize a company of Guards to help the state militia protect them. But this time the governor turned thumbs down. Meanwhile an irate "army" of two hundred Mormons started from Ohio for Missouri. But before it entered Clay County the Prophet ordered it to disband.

All negotiations for the sale or return of the Mormon lands in Jackson County came to nothing.

So many newcomers had joined the Mormon community in Clay County that now the citizens of that county in their turn began to be alarmed. They saw that it was the old story of the camel who begged to stick his head into the Arab's tent and ended by crowding out his host. Another mass meeting was held at Liberty, and the Mormons were asked to leave. It was suggested that they go to Wisconsin. Immigrants who had just come in were advised to go at once, those who owned no land as soon as the harvest was in, and landowners as soon as they could sell their fertile acres.

It was now clear to the discouraged Mormons that they would have to live by themselves, and the legislature created a new county, Caldwell County, for them in December, 1836. There the Mormons planned their new capital, which they called Far West, with a public square in the center and broad avenues. The excavation made for the foundation of the proposed temple may still be seen. Almost immediately the town had a population of some three thousand busy people. The Mormons soon occupied some two thousand farms.

Here, at last, the Mormons had political control, and early in 1838 Prophet Joseph Smith and bearded Sidney Rigdon arrived. Here they expelled some of the other leaders of the sect and, it is said, organized the Danites as a kind of Mormon armed guard.

According to the Gentiles of those days in Missouri, the Mormons took too literally the text that the Lord had "given the earth and the fullness thereof to his people," and were inclined to help themselves to the property of others. It is difficult to learn just what the facts of the matter are, since the accounts of the two parties conflict. But it appears that Sidney Rigdon, on the Fourth of July, 1838, while making a speech, declared that henceforth the Mormons would resist all invasions of their legal rights.

Had the Mormons remained in their own county, perhaps trouble might have been averted, but they overflowed into the counties around and, finding that the Whigs and the Democrats were about evenly divided in the county to the north, they sided to a man with the Democrats.

The Whigs, indignant at this, were reported to be planning to keep Mormons from voting and there was fighting on election day, though no one was killed. Immediately the Mormons invaded the county in force and there compelled a justice of the peace to give them a written pledge that he would administer the law fairly and not attach himself to any mob. This caused warrants to be issued for the arrest of Joseph Smith and several others. Public feeling ran high. Governor Boggs issued his notorious "exterminating order," declaring that "the Mormons must be treated as enemies and must be exterminated or driven from the State, if necessary, for the public peace." He called out a force of four hundred mounted men.

With their usual alacrity, the fiery Missourians sprang to arms and quickly advanced under General John B. Clark to meet the Mormon forces.

At Crooked River there was a skirmish, and a more serious fight at Haun's Mill. One of the militiamen was wounded, and a number of Mormons were killed in battle. Some were murdered after they surrendered. Their bodies were pitched into a well nearby.

When the Missourians reached Far West, Joseph Smith and five others surrendered and were held for trial. The Prophet and others were committed to jail at Liberty and severally indicted for murder, treason, burglary, arson, larceny, theft, and stealing. Having secured a change of venue, they were sent toward Columbia under guard. On the way, Joseph Smith escaped. He took refuge in Illinois. . . .

Once more the Mormons had to move out. They had already lost most of their property in the "war." They

THE MORMONS FOLLOWED THEIR PROPHET WITH

LITTLE MORE THAN THE CLOTHES ON THEIR BACKS.

followed their Prophet to Illinois with little more than the clothes on their backs.

Once more, this time in Illinois, the Mormons established their capital. The town stood on the Mississippi, and was called Nauvoo.

The Mormons next appealed to Congress to redress their grievances. Congress refused to act. But the legislature of Illinois granted a charter for a Mormon military force, the Nauvoo Legion.

By this time the whole country was aware of the Mormon troubles, and feeling against them soon developed in Illinois. It was hard to fit such a tight little theocracy into a democratic frontier state. Politicians suspected that the Mormons would vote as a bloc; merchants believed that Mormons would trade only with Mormons; the clergy attacked them from the pulpit; and they were regarded with increasing distrust and even fear. Many people believed sincerely that the Mormons, if driven out, would join the British.

At last, on June 27, 1844, Joseph Smith and his brother Hyrum were arrested. Soon after, while still prisoners, they were murdered.

This disaster at last convinced the Mormons that they must "get away from Christians" and find a home in some far-off region. Governor Ford tried in vain to protect them; he wrote that "the public mind everywhere is so decidedly hostile to them that public opinion is not inclined to do them common justice. . . . I fear they will never be able to live in peace with their neighbors. . . . Those who may think it wrong to drive out the Mormons cannot be made to fight in their defense." Ford advised the Mormons to go.

They considered a number of possible asylums: Arkansas, Oregon, California, Texas, Nebraska, Vancouver Island, the Big Basin. Brigham Young, their new leader, had to find a home quickly for thousands of impoverished people.

California seemed the most likely goal, and when the Mexican War broke out, it was arranged that a Mormon battalion should be mustered into the United States Army and so cross the plains. Meanwhile Brigham prepared to lead all his people to the west.

Brigham Young was a capable organizer. Before he started the great migration, he prepared encampments at regular intervals across Iowa from Nauvoo to Hart's Bluff on the east bank of the Missouri River. The last of these was known as Miller's Hollow, but the Mormons soon changed its name to Kanesville, after an army officer who befriended them. Some years later, when they moved out, the town was reorganized and dubbed Council Bluffs.

The first company of refugees headed west from Nauvoo in February, 1846, to be followed at regular intervals until winter set in. They suffered many hardships, having few wagons or tents. Many sickened or died along the way, what with the rain and the mud and the snow and the frost. Though the elders frowned on cardplaying and all such sinful amusements, they provided a brass band, encouraged dancing, and kept up the spirits of the migrants by appealing to their religious faith and singing songs. As he approached the Missouri, William Clayton, an English convert, composed the song "Come, Come, Ye Saints."

> We'll find the place which God for us prepared,
>   Far away in the West;
> Where none shall come to hurt, or make afraid:
>   There the Saints will be blest.
> We'll make the air with music ring—
>   Shout praises to our God and King:
> Above the rest these words we'll tell—
>   All is well! all is well!

The first company, or "camp of Israel," reached the Missouri about the middle of June. Later companies, not being able to cross the plains before winter, remained on one bank or the other of the Big Muddy. The camp on the west bank in Nebraska was called Winter Quarters and comprised a thousand log cabins before the year was out. Thousands gathered thereabouts, of whom fully three hundred died of exposure and disease.

Many Mormons took refuge among the Indians in Nebraska, but before long even the Lamanites, frightened by the destruction of game and timber on their lands, asked their guests to pull out. It seemed in those days that nobody could get along with the Saints.

The *Frontier Guardian,* for August 9, 1849, thus describes the Mormon migration from the Missouri:

"On Saturday, July 14, about noon the last wagons left Winter Quarters and began to bend their way westward over the boundless Plains that lie between us and the Valley of the Great Salt Lake. Slowly and majestically they moved along, displaying a column of upwards of 300 wagons, cattle, sheep, hogs, horses, mules, chickens, turkeys, geese, doves, goats, etc., etc., besides lots of men, women and children. In this company was the Yankee with his machinery, the Southerner with his colored attendant—the Englishman with all kinds of mechanic's tools—the farmer, the merchant, the doctor, the minister, and almost everything necessary for a settlement in a new country."

As was usual in those days, the emigrants packed too much plunder along, and used much too heavy wagons, overloading them with sawmills, blacksmith shops, gold diggers, grindstones, and heavy furniture. Most of these had to be abandoned or sold along the way. Jim Bridger made a pretty penny buying up the property of emigrants for a song.[1]

The following April, Brigham reached Salt Lake Valley and declared, "This is the place!" For years after, the Mormon migration continued, leaving six thousand dead along the way. Many of the Mormons were too poor to own a wagon or a team, but pushed and pulled their few possessions across the plains in handcarts, each person being limited to seventeen pounds of baggage. By 1860 the great trek was over.

But the Mormons never quite reconciled themselves to leaving the Missouri River; there are many of them living along its banks today, all the way up to the headwaters.

Some of the Mormons remained in Iowa and in Nebraska, and one sect, the Hedrickites, returned to Independence in 1867 and acquired title to the Temple Lot there. Another group, the Reorganized Latter Day Saints, also came back to Jackson County and tried to get possession of that site. Not to be outdone, the Utah Mormons later colonized a "stake" at Independence, so that the town long had a considerable Mormon population.

This was the first of the religious wars on the Big Muddy. It might have turned out differently, perhaps, had the Mormons been as combative as their neighbors, for they had able leaders and a strong organization. Their weakness in such a conflict lay in the fact that a great many of them were converts made abroad, who came to the western frontier fresh from Europe and lacked that long discipline of Indian fighting and private war which has contributed so much to the formation of American character. Those foreigners were simply not quick on the trigger.

CHAPTER IX

Diving for Furs

From Grand River to the Kaw—the western boundary of the state—the Missouri River was tough on navigators. It was full of shallows, narrow chutes, whole plantations of snags, where the river was "covered with wood," bars, traveling islands, drifts of logs, bends and eddies. Lewis and Clark had trouble there as Hunt and Lisa did later.

Prince Maximilian, still later, had even worse luck. The steamboat *Yellowstone*, making its second trip to the mouth of its namesake in 1833, nearly ran aground at the mouth of the Grand; was so hemmed in and bumped by snags that they had to stop the engine and pole through a channel no wider than the boat; then the engine broke; they had to stop and saw off dangerous snags; tried towing, only to snap the towrope, sending the men in a heap; ran aground and had to lighten the boat to get off; and finally were swept into a point, which tore away the whole side gallery of the boat. Apart from that, the voyage was pleasant, as the country all along the river was extremely beautiful. It still is.

The country teemed with plums, gooseberries, raspberries, wild apples, myriads of mulberries, and papaw trees offering luscious vegetable custards. Parroquets abounded, and so did sandhill cranes, pelicans, deer, raccoons, bears, mosquitoes, and skunks. Everywhere were fine tall trees, handsome prairies, and imposing bluffs. Above Grand River, among the Snake Bluffs and about Snake Island, Lewis and Clark's men heard snakes on the lake that gobbled like turkeys. The explorers failed to make note of the name of this species!

The Kaw River flows into the Missouri from the west only a few miles above Independence. To those who live along its banks, the Kaw, flowing in the same direction as the Missouri does below its mouth, seems to be part of the main stream. They claim that the Missouri from North Dakota to Kansas City is "probably the newest river" in the United States.

Geographers generally have not shared this view, but it is a fact that the Kaw had some traffic of its own.

Not far up the Kaw River was a ferry in a class by itself. Ely Moore in "The Story of Lecompton" has described it:

"The wagon boss pointed to a huge sycamore log some twenty feet long, five feet in diameter with an excavation in the center five feet in length, three feet wide and two feet deep, with a 4x6-inch scantling for a keel, remarking, 'Thar's the ferry and hyars the ferryman.' As I looked my doubts about crossing on that log, he answered my looks by saying: 'Don't feel skeery, mister, for she's as dry as a Missourian's throat and as safe as the American flag.'" [1]

The Chouteau family are a great part of the history of the Missouri River. Their fur-trading posts were scattered all up and down the main stream and its tributaries. They had one post which they reached by keelboat up the Kaw.

In March, 1829, François G. Chouteau loaded this boat, of forty or fifty tons burden, with four hundred packs of peltries billed for St. Louis. His brother Frederick was in charge. The bill of lading will give some idea of the animals whose furs formed the bulk of the trade in those days:

>   Beaver skins 65
>   Otter skins 100
>   Deer skins (20 buck and 20 doe) 40

Raccoon skins	120
Muskrat skins	500
Wolf skins	100
Badger skins	100
Buffalo robes	10

On this boat the widow, Mrs. Vasquez, and her children took passage. Her husband, Barnett Vasquez, had been the first Indian agent of the Kaw tribe. None of the children was over twelve years of age. On board also were two Indian agents, Mr. Hughes and John Dougherty, two pilots, one of them with his daughter, along with eight hands, ten Kaw Indians, and Frederick Chouteau.

The pilot, Baptiste Datchurut—an interpreter, squaw-man, and Negro freedman—was drunk and failed to show up when the boat left. Chouteau therefore employed a mulatto, Frank Zabette, to handle the heavy steering oar.

But Baptiste, hearing that the keelboat had gone without him, doused his head with water, piled into a canoe along with a discharged soldier by the name of Kennedy, and overtook the keelboat two miles above Prime's Ferry where the town of Independence is now.

The old rascal showed his papers and said, "Your brother sent me to take charge of the boat."

Frederick had no illusions about old Baptiste or his condition. But he could only say, "I am very sorry; I wish you had not overtaken us."

The wind blew hard from the south and for a time they had to lie by because of the wind. The water was high and they kept under the south bank for protection as much as they could. Near Prime's Ferry was a big rock—on the bank at low water, but now square in the middle of the racing current. Chouteau saw the danger as the boat shot down the stream, and yelled to the crew to row clear of it. They pulled manfully, but the steersman was too drunk or too old to throw the stern over. The boat crashed squarely upon the rock broadside, bashing the planks in.

The boat was then steered for the shore, though it was sinking fast. They threw out the anchor, but it dragged without catching. Chouteau and seven of the crew jumped overboard to lighten the boat and swam for the shore. But the water was cold and swift: Kennedy and two of the Canadians were swept under and drowned. The anchor caught and the boat hung precariously in the boiling river.

Chouteau hurried down to the ferry, got a flatboat, and returned to the rescue. He saved everyone, but the boat was under water with all its precious freight.

Next day the men went out in the flatboat with axes and laid alongside the sunken keelboat. They chopped a hole in the deck and fished out a few of the packs near the opening. The boat was sixty feet long and the cargo stowed from end to end below. The packs were waterlogged, deep in the drowned hold, most of them yards and yards from the opening. No one dared venture beneath the deck in that black water. If he lost his head down there or cramps took him, he might never find the opening or breathe the air again. It seemed almost as dangerous as diving through a hole in the ice.

The men stood knee-deep in the cold water on the sunken deck and looked at one another in silence. Chouteau was in despair. The whole winter's trade seemed lost.

Then the cook, a mulatto slave named Joseph Lulu, volunteered. He dived into the hold and brought up a pack of furs to be placed in the flatboat. Then he went down again. That day he dived into the hold of the sunken boat no less than 375 times, bringing up a pack of furs every time. All the furs were saved. Says Chouteau, "He was worth his weight in gold." [2]

Such a feat deserved to be remembered—and rewarded. Frederick's father freed the slave. Joseph became a fireman on a Mississippi steamboat, only to perish a few years after when the boat blew up.

The fur-bearing animals kept the boats plying up and down the river and its tributaries, but there were other creatures which played a part fully as important in the history of the valley.

There are three species particularly connected with the Missouri River and the life of the people, both white and red, who inhabited its shores. The first of these, of course, was the buffalo—the Indians' commissary. The Army repeatedly smote the Indian hip and thigh, but it is doubtful whether the small commands maintained on the frontier could ever have conquered the red men by mere force of arms. The hunters who exterminated the buffalo were the real conquerers of the Missouri. They put the Indian out of business. Having nothing to eat, he had to come in and submit.

But there were other creatures on the Missouri which very nearly put the Army out of business, and which abounded in numbers equal to those of the bison. These were the rodents—rats and mice—which swarmed everywhere along the stream and, indeed, all over the West.

Frontier posts were generally built by the soldiers who first occupied them out of logs, sod, bricks made of mud found near the site. Such buildings made an ideal home and hunting ground for rats and mice, which devoured great quantities of rations—to say nothing of their devastation of truck gardens. This was bad enough. But the rodents came near putting the army afoot, and so delayed the conquest of the Indian, probably by years.

Generally an army post was built where wood, water, and hay were available. But grain for cavalry horses had to be shipped in at huge expense, and stored in primitive warehouses which could not be made ratproof.

Though at first many of the troops sent west were

cavalry (the only troops of much use against mounted Indians), the destruction of grain by rodents caused such expense to the War Department that more and more infantry were sent to the frontier and cavalry outfits became the exception everywhere. Thus, in the campaign against Sitting Bull, Custer commanded the only considerable force of mounted men.

But the rats on shore were hardly more of a problem than the mice afloat. Since boats were tied up to the bank nearly every night, it was impossible to keep such vermin from coming aboard. The hold of every steamboat was alive with rats and mice, and smaller vessels also suffered heavy loss. Lewis and Clark, who traveled in small boats, report the loss of a considerable quantity of their goods caused by mice on the river.

Later travelers took precautions, and so brought into the picture a third species—the domestic cat. Almost every steamboat had its staff of mousers, and many carried crates of cats requisitioned by army posts or trading forts upriver. Smart settlers brought along their own toms and tabbies.

Not a few canoemen carried a cat or two on their voyages. Brackenridge tells how he was kept awake by the "doleful serenade" of some cats in a canoe belonging to passing strangers who had tied up alongside his keelboat.

He remarks that the "captain" of this party, going upriver to establish his family, had as his cargo only a barrel of whisky, some cotton for his wife to spin, and these harmless, necessary cats. And necessary they were, in those days, on the Missouri. A settler without a cat was headed for bankruptcy—bound to be eaten out of house and home.

People heading west up the Big Muddy usually disembarked near the mouth of the Kaw, for above the Kaw the course of the Missouri turned northwest. The mouth of the Kaw thus became a jumping-off place for Oregon, Santa Fe, and California, and towns sprang up thereabouts: Westport, Kansas Landing—both now part of Kansas City.

Not far from Westport were the Shawnee Indians. Various churches had missions among them. Pupils of the mission were fed and clothed and educated entirely at the expense of the church. Few of them knew any English, and no doubt most of them were orphans, "beef Christians," who came, ragged, scrofulous, and verminous, to throw themselves upon the mercy of the missionaries.

The first indignity inflicted upon a new pupil was to cut his hair short like that of an Indian in deep mourning. The second was to throw away his Indian clothing and give him and his head a good scrubbing, with a little red precipitate on his shorn scalp, supplemented by the use of a fine-toothed comb. Afterward he was furnished with an ill-fitting suit of white man's clothes and taught how to put them on and off. The children emerged from this ordeal as "shy as peacocks just plucked." Each child was given a new English name.

The schoolbooks and classwork were all in English, and in order to teach English to the Indian pupils as swiftly as possible, the missionaries forbade their pupils to speak the Shawnee tongue except when absolutely necessary. This rule had curious results.

Dr. Barker, superintendent of the Shawnee Baptist Mission, went zealously to work, burning the midnight oil, month after month, translating the New Testament into the Shawnee tongue. That done, the indefatigable evangelist printed the book himself on a hand press and distributed copies to all the Indians. Only then did he discover that only Indians who had gone to school could read, and that those who had gone to school could read only English.

Some strange communities came up the Missouri to settle unoccupied lands. Among these was a society known as the Vegetarian Kansas Emigration Company. This was projected by Henry S. Clubb in 1855 in order to establish a permanent home for vegetarians. Just why vegetarians were more in need of a permanent home than other mortals was explained in the glowing prospectus of the organization. Vegetarians of common interests and aims were to be brought together, since otherwise they, "solitary and alone in their vegetarian practice, might *sink* into flesh-eating habits."

For some reason the region about the mouth of the Kaw attracted enthusiasts of all sorts. There, at the crossroads of the continent, they swarmed and settled. The valley became, for a time, the home of idealists, martyrs, and crusaders, and also of fanatics, rebels, and bandits. Call them what you please, they were all ready to fight for their beliefs.

The most celebrated of these enthusiasts on the river was John Brown.

CHAPTER X

Hell on the Border

FROM the mouth of the Kaw upward, the Big Muddy was not only a trail but a boundary—the boundary between the woodlands, lakes, and farms on the east bank and the buffalo plains and ranches on the west. Today on Snake Butte, on the banks of the Missouri not far from Pierre, South Dakota, stands the Center Monument, an obelisk marking the approximate center of North America. Thus the river is the great divide, where the West begins, a social barrier between two cultures, two climates, two ways of life. Where it flowed between Kansas and Missouri, it was a political and even a military boundary.

Every man, they say, is entitled to one war, but the people on the Missouri River about the mouth of the Kaw have had more than their share. There the Civil War began years before the attack on Fort Sumter, and there it dragged on long after Lee's surrender.

The people along the river were not themselves primarily to blame for the conflict. Rather, the blame falls on Congress. If that body had gone seriously and honestly to work, the slavery question might conceivably have been settled in a peaceful manner by economic or political means, and the Civil War might have been avoided.

But as tension over slavery continually mounted in the States, Congress sidestepped the issue and passed the buck to the settlers on the frontier.

The law known as the Missouri Compromise had forever

prohibited slavery in all the territory of the Louisiana Purchase north of 36° and 30' *except* the state of Missouri. This left Missouri sticking up like a sore thumb into free territory. Slaves had always been held in Missouri; and since the treaty with France expressly provided for the protection of the property rights of the inhabitants, Missouri was a slave state.

But in 1854 Congress passed the Kansas-Nebraska Act. This set up two territories—Kansas and Nebraska—west of the Missouri River, and provided that settlers in the two new territories should decide for themselves whether or not these should be slave or free. The fat was in the fire.

For the new law amounted to an invitation to both free-soil and proslavery men to swarm into the new lands and fight it out—and this they proceeded to do without delay. Their struggle falls naturally into three stages; prewar, war, and postwar. But it was *all* war to the settlers, and war of a peculiarly inhuman and savage kind.

Because the slave interest controlled the Southern states, and the South dominated Congress, slaveholders in Missouri naturally assumed that the Kansas-Nebraska Act had been passed for their benefit and with the express purpose of extending slave territory. Some were willing to permit Nebraska to be settled by free-soilers, but all felt it imperative that Kansas, lying just to the west of Missouri, should be slave territory. Already Missouri was flanked on the east and the north by free soil into which slaves were continually escaping—often with the aid of the Underground Railroad run by Abolitionists. If Kansas, to the west, should be free soil also, Missouri would find herself spang in the middle of the "nigger stealers," and slaveholders might find their investments altogether too risky.

Accordingly, great numbers of proslavery men swarmed over into Kansas and pre-empted as much land as possible. They started the towns of Kickapoo, Atchison, and Leavenworth on the Missouri River. Few of them, however, made any improvements on their claims or took the legal steps necessary to acquire title.

Meanwhile, in the North, antislavery sentiment was outraged by the Kansas-Nebraska Act, and before long the New England Emigrant Aid Company was organized to send free-soil settlers into Kansas. They came in increasing numbers and, having come from so far away, were for the most part bona fide settlers who built homes on their farms and soon established half a dozen towns west of Kansas City. Their capital was Lawrence.

Proslavery men then also organized various associations, and on election day thousands of them crossed the border in armed bands, marching with music and flags to stuff the Kansas ballot boxes. In March, 1855, though the census showed less than three thousand voters in Kansas, more than six thousand votes were cast. For men who set such store by their legal rights, the fellows from Missouri were all too ready to override the law.

Most settlers from the North shared the sentiments of Abraham Lincoln as to the Kansas-Nebraska Act. He wrote: "I look upon that enactment not as a law, but as a violence from the beginning. It was conceived in violence, is maintained in violence, and is being executed in violence. I say it was conceived in violence, because the destruction of the Missouri Compromise, under the circumstances, was nothing less than violence. It was passed in violence, because it could not have passed at all but for the votes of many members in violation of the known will of their constituents. It is maintained in violence, because the elections since clearly demand its repeal; and the demand is openly disregarded."

In that letter to his Southern friend, Joshua F. Speed, Lincoln went on to say: "You say that if Kansas fairly votes herself a free State, as a Christian you will rejoice at it. All decent slaveholders talk that way, and I do not doubt their candor. But they never vote that way. Although in a private letter or conversation you will express your preference that Kansas shall be free, you would vote for no man for Congress who would say the same thing publicly. No such man could be elected from any district in a slave State. You think Stringfellow and company ought to be hung; and yet at the next presidential election you will vote for the exact type and representative of Stringfellow. The slave-breeders and slave-traders are a small, odious, and detested class among you; and yet in politics they dictate the course of all of you, and are as completely your masters as you are the master of your own negroes. . . . As a nation we began by declaring that 'all men are created equal.' We now practically read it 'all men are created equal, except negroes.' When the Know-nothings get control, it will read 'all men are created equal, except negroes and foreigners and Catholics.' When it comes to this, I shall prefer emigrating to some country where they make no pretense of loving liberty,—to Russia, for instance, where despotism can be taken pure, and without the base alloy of hypocrisy."

It was the old question of human rights as against property rights, which has agitated our country since the days of Thomas Jefferson.

In prosperous Missouri, as in most other parts of the South, politics was largely in the hands of lawyers and rich landowners—men who had a strong sense of property and stood firmly on their legal rights. In Kansas, on the other hand, the influential people were preachers, teachers, and small farmers, who put "moral" and "natural" rights first, legal and property rights second. The Missourians, for the most part, had settled in the state because of the opportunities it offered. Missouri was then a land of rich harvests of hemp, tobacco, cotton—all crops adapted to slave labor. The Kansans were newly settled on small, bare, prairie farms in what was then called the Great American Desert. Many were colonists frankly sent out and financed by Abolitionists in New England and elsewhere, animated by a fiery love of freedom and a determination that slavery, which they abhorred, should never spread to their communities. They came in the spirit of "The Battle-Hymn of the Republic":

> As He died to make men holy,
> Let us die to make men free.

Most of the great religions of the world have originated in semiarid lands, where men are much in the open and can look up at the stars. The motto on the Great Seal of Kansas is *Ad Astra Per Aspera.*

Such men, in such a country, were bound to be passionate, if not fanatical, in pursuing their ideals. Indeed, idealism is notably characteristic of the whole Plains region, which has produced or attracted leaders of many programs called radical in their time: Abolition, Populism, Mormonism, Prohibition, Free Silver, the Nonpartisan League, Woman Suffrage, and a host of other religious, political, and social movements, including the Indian Ghost Dance.

The plainsman has the habit of looking in four directions and always stands squarely at the center of the visible world. He is, therefore, more individualistic and perhaps more egoistic than his brothers in the mountains or the forests. He seldom gives in and almost never gives out. His vision of the Plains is likely to be that of the newcomer who, standing on the bluffs above the Missouri, stared in silence and then shouted "Oceans of glory!"

So, although the people of the Missouri Valley have pretty consistently stoned their prophets at first, the idealists have nearly always prevailed there in the long run. A multitude of reforms—or changes, if you prefer—begun or developed there were later accepted by the whole country and are now part of the American Way.

"Through hardships to the stars" well expresses the history of those early Kansans. They had a hard row to hoe and many died in the furrow. What of that? Their bodies might lie moldering in the grave, but their souls went marching on.

The success of their programs was of course partly due to our national idealism and democracy, our national tradition of rebellion against oppression. But it was also partly due to the plain fact that your practical man fights only when his interests are involved, whereas your idealist fights on forever. When he falls fighting, his spirit and his cause may live on; but when a practical man dies, he is dead all over.

Idealists, of course, often find themselves unpopular. They go about stepping on other people's toes, and few men can see beyond the end of their toes.

Though the Kansans were convinced that (to quote Lincoln again) "if slavery is not wrong, nothing is wrong," they were not at first inclined to resist the highhanded Missourians with force. They had come to establish a free state by legal means, and this they proceeded to do by soberly holding an election of their own and setting up their own separate territorial government.

The fire-breathing men from Missouri made the mistake of underestimating their unresisting opponents. Having licked the Indians, bluffed the Mormons, and overwhelmed the Mexicans in a series of easy victories, they fancied themselves invincible. Few of them had ever run head-on into the New England conscience. Accordingly they mouthed a great many violent threats and, spurred on by the fiery oratory of Senator David Atchison, prepared to march in and attack Lawrence.

The news of this threatened violence soon reached the North, and the men who were backing the free-soil settlement in Kansas redoubled their efforts and raised thousands of dollars with which to purchase guns and ammunition for the defense of free-soil homes. In that day the Sharps breech-loading carbine had proved its deadly effectiveness and its superiority over the muzzle-loading rifles of the backwoodsmen. Accordingly, the Abolitionists bought hundreds of these Sharps carbines and shipped them off to Kansas.

One of the most militant of the Abolitionists was a minister in Brooklyn, Henry Ward Beecher. He contributed heavily towards arming the Kansans, and the New York *Tribune* of February 8, 1856, quotes him as saying that "he believed that the Sharps Rifle was a truly moral agency, and that there was more moral power in one of those instruments so far as the slave-holders were concerned than in a hundred Bibles."

Knowing that the proslavery men were lying in wait to prevent arms reaching Kansas, the Sharps carbines were sometimes packed in cases marked "Bibles," and because of Beecher's declaration (quoted above) were soon known as "Beecher's Bibles." In the conflict which followed, these "Bibles" were decisive factors.

In 1856 a shipment of Sharps carbines was made to Kansas via the steamboat *Arabia*. The steamboats on the Missouri were owned and operated for the most part by proslavery men, and it did not take them long to find out what was in those packing cases. The news soon spread, and when the boat touched at Lexington, a gang of men boarded the steamer and carried them off.

But when they broke open the cases and examined the guns, they found that all the breechblocks had been removed, rendering the weapons useless. Dr. Calvin Cutter had anticipated the possibility of such a raid and had sent the breechblocks to Kansas overland. The baffled Missourians cussed the whole business as a "Dam-Yankee trick."

All along the border, the murders and killings went on. It was clear that open war could not long be postponed.

The first attempt to attack the Free-Soil Capital in

JOHN
BROWN

Kansas was called off, but in 1856 a force of proslavery men, by that time known to the world as Border Ruffians, actually marched on Lawrence, destroyed the newspaper presses, burned the hotel, looted the town, and killed three men. Things looked black to the Free-Soil party. Lincoln himself expected Kansas to come into the Union as a slave state; but at that time Lincoln had never heard of John Brown.

Tall, gaunt, nearing sixty, with burning blue eyes and a jaw like a steel trap, John Brown was decisive, fearless, urgent, without compromise for evil, ready to destroy the sinner with his sin. Already he had helped dozens of slaves escape to Canada. He carried a Sharps rifle in one hand and a Bible in the other, and wore a bowie knife in his belt. In fact, he loved cold steel, and was forever sharpening cutlasses and making pikes with which to arm the slaves. He had come to Kansas on purpose to fight for freedom, to stamp out slavery; and though, like other poor men, he had to make a living, he had little personal ambition other than to make his will, which he believed to be the will of God, prevail.

John Brown had five huge sons, all ready to fight and die with him, and no lack of other followers—to whom the Missourians were simple "mobocrats" and "slavers."

He knew his Bible by heart and in it found texts that inflamed his followers: "An eye for an eye; a tooth for a tooth. . . . Because I delivered the poor that cried, and the fatherless, and him that had none to help him: I put on righteousness and it clothed me. . . . I brake the jaws of the wicked and plucked the spoil out of his teeth. . . . Fear not them which kill the body, but are not able to kill the soul."

Having learned of the proposed raid on Lawrence, John Brown led his little band to the rescue. He came too late. From afar he saw the smoke of the burning town. Apparently that decided him. That night he and his friends visited the cabins of certain proslavery men, dragged five of them from their beds, and chopped them down with sabers and lead. John Brown gave his enemies a taste of their own medicine. He taught the Kansans to fight back.

Afterward things became so bad that in 1857 federal troops were sent in. A fair election was held, and the Free-Soilers carried the territory. Strife in Kansas was at an end.

But strife along the river continued—this time on the *other* side of the border. This time it was men from Kansas who rode by night into Missouri, carrying off Negroes and stock and killing settlers. These riders were called Jayhawkers and they soon chased a good many of the Border Ruffians downriver. John Brown himself carried off eleven Negroes in one night. The governors of Kansas and Missouri joined in patrolling the border in the interests of peace, but all in vain. The Jayhawkers continued their raids, and Freebooters from Kansas, encouraged by Senator James H. Lane, looted and burned towns on the Missouri side of the river.

Having taught the Kansans to fight, Brown carried his war to Harpers Ferry, Virginia. There in 1859 he seized the arsenal, planning an insurrection of the slaves. It seemed a hopeless gesture, but it cut the Gordian knot. Brown's action in Kansas had set the frontier aflame. His action at Harpers Ferry threw all men in the States to one side or the other.

By a strange coincidence the officer who captured Brown was Robert E. Lee.

Condemned to hang, Brown said, "I will show men how to die for the truth." After John Brown's body was dead, the presiding official declaimed, "So perish . . . all such foes of the human race." But it was not so simple as all that. John Brown's hanging brought the Civil War from the frontier to the States.

In Missouri, as in other border states, opinion was sharply divided on the question of secession. But much to the surprise and dismay of Southern sympathizers, the state voted to remain in the Union. Some forty thousand Missourians enlisted in the Confederate forces; but nearly three times that many fought for the Union.

After Fort Sumter was fired on, affairs on the Kansas-Missouri border went from bad to worse. A great many engagements were fought in the state of Missouri (most of them south of the river) until in the battle of Westport—that decisive "Gettysburg of the West"—the Confederate forces were crushed. When it was all over, Missouri abolished slavery almost a full year before the nation adopted the Thirteenth Amendment.

But all this while the warfare between guerrillas on the border had continued with mounting fury. The Kansans, beginning in 1860, looted and burned one town after another in Missouri. Red Legs from Kansas (of whom Buffalo Bill is said to have been one)—so-called because they wore red morocco leather leggings—kept busy raiding Missouri border counties.[1]

The Missourians on their part were not idle. Under the leadership of the ferocious Bill Anderson and the bloodthirsty William Clarke Quantrill, they swept into Kansas, killing and looting. Bill Anderson was ruthless. It is said he carried the scalps of two white women dangling from his bridle bit. He did not hesitate to murder whole batches of prisoners of war.

Quantrill, whose application for a commission in the Confederate forces had been rejected by the Confederate secretary of war because of the man's bloodthirstiness, was nevertheless commissioned by General Sterling Price on the frontier. Somewhat more than one year later, August 21, 1863, Quantrill, leading 250 Missourians from Jackson County, swept into Lawrence giving orders to "kill every man, burn every house." He sacked the town. Almost two hundred people were killed.

In Quantrill we see the transition from the zealot to the bandit. For all his bloodthirsty talk, he was thoroughly mercenary and could, on occasion, be bought off. The laudatory border ballad which bears his name inadvertently gives away this side of his character.

> Oh, Quantrill's a fighter, a bold-hearted boy,
> A brave man or woman he'll never annoy,
> He'll take from the wealthy and give to the poor,
> For brave men there's never a bolt to his door.[2]

He was killed while running from his enemies in Kentucky in 1865.

But while he rampaged on the Missouri border, he became a legend and remained a thorn in the flesh of the Kansans. His exploits, in fact, caused General Thomas Ewing of the Eleventh Kansas Infantry Volunteer Regiment to issue his notorious Order Number Eleven. Ewing acted under the threat of Senator Lane of Kansas that he would be a "dead dog" (be cashiered) if he failed to issue it. This order required all persons in the border counties immediately south of the Missouri River, except those living within one mile of the limits of one of the principal towns, to remove within fifteen days. Loyal men could stay at military stations or retire far into Kansas. All grain and hay was to be delivered to the military posts.

This order compelling Missourians to move out worked great hardship and was violently resented. The Jayhawkers had a field day, burning the prairies and buildings throughout what was afterward called the Burnt District. Three counties were depopulated. No doubt the Mormons, then safe in Utah, who had been driven from those very counties by the Missourians, heard the news of their removal with a certain grim satisfaction. The artist, Bingham, then a colonel on Ewing's staff, afterward painted a celebrated picture showing the distress of the Missourians, entitled "Order Number Eleven."

After the Civil War had ended in other parts of the country, it still went on along the western border. The guerrillas, accustomed to living on the country, used to bribing friends and robbing enemies, seemed unable to stop. Their feeling was, in the words of the ballad:

> Oh, I'm a good old rebel, that's what I am,
> I won't be reconstructed and I don't give a damn.

Men of this sort had little use for law and order, and none at all for Yankee law and order. They infested Missouri, Kansas, Arkansas, and Indian Territory (now Oklahoma), perpetuating a tradition of banditry that has not entirely evaporated even yet.

Of this, the third stage of the war on the Missouri, Jesse James is the representative figure. His exploits as a train robber and bandit are famous. After avoiding capture for nearly sixteen years, he finally settled in St. Joseph under the name of Thomas Howard. He was regarded with a sort of sneaking pride even by those who disapproved of his larcenous activities, for he was a man of some originality, having invented the train robbery. This first railroad stickup

THE JAMES BOYS AND THEIR GANG ROBBE

D PLUNDERED HALF A DOZEN STATES.

took place at Adair, Iowa, July 21, 1873. In such hard times, when the railroads' attitude was "the public be damned," train robbers were regarded by multitudes of law-abiding people with what can only be described as friendly tolerance.

Of course Jesse, having been a guerrilla, did not find it easy to "go straight." The guerrillas had been especially cruel to Union men in Missouri. These now, having the law on their side, were ready enough to retaliate. An honest man can get along, if necessary, with few friends, but a bandit must have many friends or perish. So it came about that Jesse James acquired the reputation of a Robin Hood. The James boys and their gang robbed and plundered half a dozen states, but particularly their own. Unhappily for them, they robbed Southern and Northern people too impartially. As times got better, railroads and banks became more popular, and the law chased Jesse out of Jackson County.

With Jesse in his little white house with green shutters in St. Joseph lived his family and two new members of his gang, the Ford brothers—Charles and Bob—who joined Jesse in 1882. The governor had put a price of $10,000 on Jesse's head.

But Jesse was nothing if not respectable—when not engaged in his nefarious exploits. Blue-eyed, bearded, and soft-spoken, he made no trouble, and few citizens of St. Joseph knew Thomas Howard for the outlaw Jesse James.

On the morning of April 3, 1882, Jesse walked into the room where the two Ford brothers were activating their rocking chairs, complained of the warm weather, and removed his coat. Jesse wanted to go out into the yard, but feared someone would see his pistols. So he unbuckled his belt and laid his guns on the bed.

Then, instead of going outdoors, he paused. His eye was caught by a picture on the wall. It was a favorite of his—and it was dusty. He picked up a feather duster and stepped up on a chair to clean it. The two men behind him rose swiftly and jerked their guns. Bob Ford shot Jesse through the back of the head.

The Ford brothers started to leave, but Mrs. James called them back and accused them of the murder. The neighbors swarmed in. It was some time before people could believe the dead man was actually the famous bandit. The Ford brothers were arrested and pleaded guilty to the murder.

Meanwhile, almost before Jesse was buried, balladmakers took their pens in hand:

JESSE JAMES

Jesse James was a lad that killed a-many a man;
He robbed the Danville train.
But that dirty little coward that shot Mister Howard,
Has laid poor Jesse in his grave.

Chorus:
   Poor Jesse had a wife to mourn for his life,
   Three children, they were brave.
   But that dirty little coward that shot Mister Howard,
   Has laid poor Jesse in his grave.

It was Robert Ford, that dirty little coward,
I wonder how does he feel.
For he ate of Jesse's bread, and he slept in Jesse's bed,
Then laid poor Jesse in his grave.

Jesse was a man, a friend to the poor,
He never would see a man suffer pain;
And with his brother Frank he robbed the Chicago bank,
And stopped the Glendale train. . . .

The people held their breath when they heard of Jesse's death,
And wondered how he ever came to die;
It was one of the gang called little Robert Ford,
He shot poor Jesse on the sly.

Jesse went to his rest with his hand upon his breast;
The devil will be upon his knee.
He was born one day in the county of Clay
And came from a solitary race.

Jesse went down to the City of Hell,
Thinking for to do as he pleased;
But when Jesse come down to the City of Hell,
The Devil quickly had him on his knees.

This song was made by Billy Gashade,
As soon as the news did arrive;
He said there was no man with the law in his hand,
Who could take Jesse James when alive.

Chorus:
 Poor Jesse had a wife to mourn for his life,
 Three children, they were brave.
 But that dirty little coward that shot Mister Howard
 Has laid poor Jesse in his grave.

 The Ford brothers were sentenced to be hanged within forty days; but to the surprise of everyone, they just grinned at the judge and strolled back to their cells, quite unabashed. Next day their pardons came from the governor, who had sent them to St. Joseph to get Jesse.

 In that city now stands the Jesse James Hotel. The management advertised with the slogan: "You won't be held up at the Jesse James." Not long ago some bandits stuck up the hotel bar and skipped with the cash. Apparently there is humor among thieves.

CHAPTER XI

Kaycee

IN KANSAS CITY also they named a hotel James—not after Jesse, but after the saint. At that St. James Hotel a significant incident occurred. Ever since the days of the free range, Kansas City, now the biggest cattle market in the world, has been a rendezvous for cowmen. One day Shanghai Pierce, the gigantic Texas cattle king with the foghorn voice, came to the hotel and called for his mail. A new clerk stood at the desk. Not knowing Pierce, he looked up and said, "Who are you?"

Pierce inhaled, glaring at the poor ignoramus for a minute, then bellowed, "I'm Shanghai Pierce, Webster on cattle, by God!"

This story is significant as well as amusing, for the reason that even the arrogant Shanghai Pierce felt called upon to make some claim to culture in Kansas City. For this town at the mouth of the Kaw has always had a certain air about it, a certain frontier love of the best which went with the character of the daring young men who, generation after generation, have made it their headquarters.

Too many people have a notion that all the frontiersmen were plodding clodhoppers, goading oxen across the plains. But the men who made Kansas City their outfitting point were seldom of that description. Indians, explorers, army officers, trappers, traders, buffalo hunters, cattlemen, steamboat pilots, railroad magnates, industrialists, and financiers—all had a certain flair for manly elegance—even if it went no further than a love of fine feathers and a readiness to fight. Here at the bend of the Missouri, North, South, and West met. The result is a city, which, with all its industrial and financial advantages, is more remarkable for a certain vigor, briskness, and enthusiasm for the good life.

It was so in the days of Kit Carson and William Bent, in the later times of Custer and Buffalo Bill. It was so when Market Square was the summer camp of the buffalo hunters, of Wyatt Earp, Billy Dixon, Wild Bill Hickok, and Jack Gallagher—who paraded in boots of fine black calfskin, fancy vests, white linen, black string ties, and long-tailed frock coats of black broadcloth with velvet collars. And it was so when the cattlemen made their drives overland or rode the caboose of a train full of fat steers into Kaycee.

Today, accordingly, Kansas City is a town of fine buildings, handsome monuments, beautiful parks, broad boulevards, libraries, museums, good hotels, and pleasant homes.

It was never a city of angels, and has had its gang wars and corrupt political bosses. But since 1880, Kansas City has been the home, one might almost say the mate, of a great newspaper—the *Star*.

The paper was founded by William Rockhill Nelson, a man of immense energy, plenty of brains, and a public spirit fully as strong as his marked abilities. He had his own theories about running a newspaper, regarding the reporter as the essential man on his staff. He demanded much of his many reporters and insisted that they be far more than errand boys for the editors. He set out not merely to tell the truth, but to be the mentor of his town.

He looked after his patrons and their interests with unceasing vigilance. He remained in the harness for thirty-five years, making endless war against municipal fraud, monopoly, gambling, and vice—an endless campaign for the improvement of his city. Some people considered him a busybody because he went around treading on other people's toes. But he trod so hard that they generally changed their position before he got through. Though he sometimes lost battles, he generally won his wars. His paper became a local and regional influence without parallel in journalism along the Missouri River.

As to his credentials, it might be enough simply to say that he won the admiration of the late William Allen White, who was a reporter on the *Star* in the nineties. White described Nelson in *Collier's*, June 26, 1915, in these words: "He was big—monumental, with a general Himalayan effect as he appeared behind his desk. He had a great voice; in his emotional moments—which were not infrequent—this great voice rattled like artillery." White went on to say that Nelson was "a ruddy-faced, square-shouldered, great-bodied, short-legged man."

Nelson was the Nestor of Kansas City, and when accused of being too nosy and meddlesome, defended himself as follows:

"Under the malign direction of Nelson, the *Star* has kept things constantly stirred up. It has made tenants dissatisfied. They never used to complain about light and air. Now they won't look at a house unless every window opens on a flower garden with a humming-bird in it. The *Star* won't let anybody alone. It insists on regulating the minutest detail of people's lives. Its regulations are pernicious and extravagant. Its preaching about more parks and boulevards and breathing spaces and supervised playgrounds for children, and Dorothy Perkins roses, and swat the fly, and housing reforms, and a new charter, and art galleries, and keep your lawn trimmed, and take a lot of baths, and throw out the bosses, and use the river, and cut the weeds on vacant lots, and read the Home University Library, and for God's sake don't build such ugly houses, and make the landlord cut a window in the bathroom, and put goats in Swope Park, and why mothers risk their babies' lives by bringing them up on bottles, and plant your bulbs now, and teach your children manners, and what's the use of lawyers, and cultivate a pleasant speaking voice, and build a civic center, and put out houses for the birds, and walk two miles before breakfast, and why are Pullman cars so hot in winter, and go to church, and cut out the children's adenoids, and build trafficways, and the square deal, and sleep with your windows open, and smash the saloons, and pooh-pooh on factories that employ women, and reduce the street-car fares, and use two-by-sixes instead of two-by-fours if you want your house to stand up, and move out in the suburbs, and build hard-surface roads everywhere, and all the other things, have increased the cost of living and given people inflated ideas, and pretty nearly ruined the town." [1]

Nelson fought for the man, not for the party; and when he had made up his mind, never bothered to ask whether his stand would be popular. Still, he was generally progressive and, after Theodore Roosevelt left the White House, engaged the ex-president as a contributing editor of the Kansas City *Star*. Nelson left his fortune to found a splendid art gallery for his city, and after the death of his heirs, his paper was sold at auction so that the proceeds might go to the gallery.

The editors and managers of the *Star* and of his other paper, the *Times*, promptly organized a stock company of the employees and so raised enough cash to buy it.

Everybody expected that a paper in which all stockholders were employees would go on the rocks in short order, but nothing of the kind has happened. The *Star* continues to be prosperous and public-spirited, and recently had no small part in ousting the notorious Pendergast regime.

But, somehow, everything in Kansas City offers some cultural advantages. Under Prohibition, when Kansas City was "wide open," Negro bands, not permitted in the better places, were made welcome in the dives and joints. Negro musicians, composers, and arrangers swarmed in from the South and West.

All through the twenties, conditions in Kansas City were lush and inviting. Musicians found more work than they could handle. They soon developed a Kansas City "style" which has perhaps been best defined by Count Basie. As he put it, "I don't dig that two-beat jive the New Orleans cats play, because my boys and I got to have four heavy beats to a bar and no cheating." [2]

The Kaycee jazz musicians have a tradition peculiar to themselves, which calls for tuning their instruments when they start playing. Their patron saints are the late Bennie Moten and George Ewing Lee. Lee learned to play in the United States Army during World War I and came back to organize a band in his home town, Kansas City. He soon became nationally known. A whole flock of band leaders and performers have sprung to fame in that "Heart of America" city on the rocky bluffs overlooking the Missouri. There jazz music matured into big business.

"Musicians were everywhere. Pete Johnson, pianist, and Joe Turner, blues shouter, teamed up and were at their best at 4 o'clock in the morning when musicians got off work and dropped around to gulp beer from large tin cans selling for a nickel—pay when served. Hot crawfish, chittlins, barbecued ribs and dime-a-drink whisky were on tap at all times. Everyone had money, and no one tried to save it. The town was jumping. Jazz was king." [3]

Those who remember those days say that it was the custom of Count Basie, in the wee small hours of the morning after the regular patrons of the club where he played had gone home, to play for the musicians who dropped in after their jobs were done in other places.

The musicians' Local Union #627, under the leadership of William Shaw, throve mightily and prosperity continued until the mid-thirties. Then the repeal of Prohibition cut down the numbers of free spenders visiting the town. But the driving, riff-filled Kaycee brand of jazz remains an important contribution to music on the Missouri. Even today, long after the peak of 1937, there are plenty of night spots around Eighteenth and Vine where small bands play fast or slow blues, using a piano, drums, a sax, bass, and trumpet.

Fortunately many of the best performances of boogie-woogie artists are found on phonograph records. The constant improvisations of these musicians make it unlikely that a casual visitor will hear a first-class performance from such a band. Unlike classical music, jazz may usually be best heard from a record.

Of course there must have been many, many songs sung by rivermen on the Missouri, but, as Mr. John A Lomax has pointed out in the *Southwest Review*, "The fact is that very little material in the field of river folk song is available in print, and it is in folk song that one could have most easily seen the life of the people on the river . . . in the days when our rivers were busy with canoes, rafts, flatboats, barges and steamers there probably were many songs that had a real function in the life of river people in the same way that the cowboy songs and the songs of the sailors functioned in the life of these folk; but few of these old river songs have turned up in collections; their echoes have drifted away down the quiet streams and are now lost in the sea of the past." [4]

One such song, "The Sioux Indians," though the adventure it narrates took place on the Platte, makes mention of the Missouri River. Another, "The Wide Missouri," given in *Songs of the Rivers of America*, edited by Carl Carmer for this series, was long ago adopted—and adapted—by the United States Field Artillery and, when I wore the crossed cannon on my collar, was second in popularity only to "The Caissons Go Rolling Along."

In the time of the Indian wars and after, artillerymen

evidently felt, after looking at the Missouri, that the adjective "wide" was quite inadequate to describe that mighty, monstrous stream. They changed the word to "wild"—which was certainly more appropriate.

In those days, when our field artillery was horse-drawn, there was naturally some rivalry with the other mounted service, the cavalry. In the old days the cavalry enlisted colored soldiers, whereas the field artillery consisted entirely of white men. This fact inspired the artilleryman's version of the song:

> I loved a girl, her name was Nancy.
> Hi, ho, that rolling river.
> She would not have me for her lover.
> I am bound for the wild Missouri.
>
> She said she loved a cavalry soldier,
> Hi, ho, that rolling river.
> She would not have me for her lover;
> I am bound for the wild Missouri.
>
> Then she had a nigger baby.
> Hi, ho, that rolling river.
> She would not have me for her lover;
> I am bound for the wild Missouri.
>
> It must have been a cavalry soldier.
> Hi, ho, that rolling river.
> She would not have me for her lover;
> I am bound for the wild Missouri.

No doubt many another popular ballad was altered in sentiment and content in like manner as the great migration westward followed up or crossed the Big Muddy. The Missouri altered many things besides its banks. The people who lived with it were changed. The more we understand their strange and passionate stream, the better we shall understand, admire, and love them.

CHAPTER XII

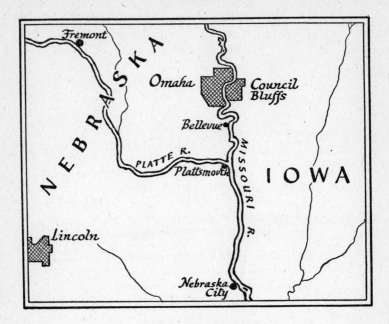

Ak-Sar-Ben

Says I, "My dearest Sally, oh, Sally, for your sake,
I'll go to Californy and try to raise a stake."
Says she to me: "Joe Bowers, oh, you are the chap to win.
Give me a kiss to seal the bargain." And she threw a dozen in.

From the mouth of the Missouri clear up to Great Falls, the stream has always been a chain of points of distribution of both goods and men. From every settlement and steamboat landing trails led and still lead to the west, trails over which travelers have streamed on rails, in planes, and motorcars, in trucks and covered wagons, in saddles and jouncing stagecoaches, in windwagons with all sails set or by shank's mare. From St. Charles and Franklin, from Independence and Westport, from Atchison, St. Joseph, Leavenworth, Omaha, and Council Bluffs, from Sioux City, Yankton, Pierre, Mandan, and Bismarck, from Fort Union, Fort Buford, Fort Benton, and from many another settlement, away they went with plodding pack trains, lumbering ox teams, and carts. The Great Plains were laced with roads following up the tributaries of the Big Muddy to the Rockies and the Coast.

They went to trade, they went to dig gold, they went looking for free land. For as soon as one gold rush or one migration ended another began. In our own time we have seen thousands of migrants traveling in old flivvers, pushing baby carts, or hitchhiking to the Coast from the Missouri.

In early days several large companies were organized to handle the business. One was the Hollady Overland Stage Line, whose owner and manager earned the tribute Mark Twain paid him when he wrote, on learning that Moses had required forty years to lead the Hebrew children from Egypt to the Promised Land, that Hollady would have got them through in forty hours.

The Butterfield Overland Mail Line left the Missouri at Tipton, heading southwest. But the biggest and best-known company with headquarters on the Missouri was the firm of Russell, Majors, and Waddell, who had offices on the east bank at St. Joseph and field headquarters near Fort Leavenworth on the west bank. The firm at one time employed six thousand teamsters, owned forty thousand oxen, had wagon parks covering acres of ground, with huge mounds of spare parts for their wagons. They transported millions of pounds of freight every month across the Plains.

William Bradford Waddell, the most imaginative of the partners, was a strict Christian, familiarly known to his employees as "Bible Bill." He made each of his men own and carry a Bible and take an oath when employed never to work on Sunday, never to drink or swear.

This regulation of Bible Bill's put a severe strain on his bullwhackers and mule skinners, for it was firmly believed on the Missouri that oxen, like roustabouts, could not be managed efficiently without considerable profanity:

> I pop my whip, I bring the blood,
> I make my leaders take the mud,
> We grab the wheels and turn them round,
> One long, long pull, we're on hard ground . . .
>
> When I got there the hills were steep,
> 'Twould make any tender-hearted person weep
> To hear me cuss and pop my whip,
> To see my oxen pull and slip.[1]

One day, according to Paul I. Wellman, Waddell caught one of his bullwhackers swearing at his team. He rebuked the man.

The bullwhacker shifted his cud to the other side of his mouth, pushed up his hat brim and offered this defense:

"Boss, the trouble with them oxens is that they don't understand the kind of language we're talkin' to 'em. Plain 'gee' and 'haw' ain't enough under the present circumstances. Now, if you could jest find it convenient to go off on that thar hill, somewhere, so's you couldn't hear what was goin' on, I'd undertake to get them oxens out."

Waddell had a sense of humor and also a sense of the practical. He walked away out of hearing.[2]

Nevertheless, Waddell persisted in his attempt to keep his organization clean, and when, in 1860, the Pony Express was organized by the firm, the riders were required to take the same oath as the bullwhackers.

The Pony Express was the Northern answer to the Southern stage line to the Coast owned by Butterfield.

An expensive venture, undertaken for patriotic reasons, the Pony Express continued for only seventeen months and bankrupted the company. But Waddell never complained. His riders, he felt, had kept California in the Union, and had, besides, made an inspiring record by carrying the mail through from St. Joseph to Sacramento in seven days and seventeen hours when carrying the inaugural address of President Lincoln to the Coast. The first rider started on a coal-black horse from the Pike's Peak stables in St. Joseph, dashed down to the Missouri, crossed the river on a ferryboat, and galloped away to the West on relays of horses until he could hand his precious mochila [3] to the next rider to speed it on its way.

Though the Missouri River was a busy highway and jumping-off place for people heading west, it had always been a well-peopled stream. Many, having come to the jumping-off place, took one look at that fertile valley and refused to jump. Instead, they stayed on its banks and for one reason or another built their Indian villages, trading forts, government agencies, missions, military posts, their mining camps, towns and cities.

One hot August afternoon a tall, lank, sad-eyed lawyer from Illinois came onto the shady tavern porch at Council Bluffs, Iowa, took off his stovepipe hat, pulled out a bandanna, and wiped his perspiring brow. He was a stranger there. Some lots had been offered him as security for a personal loan, and he had gone out on the bluff to see the property. Like everyone else who stands on a bluff above the Missouri River, he looked across. There he saw the village on the Nebraska side—Omaha.

In 1859 a transcontinental railway was the glittering dream of the West. Before long the tall, perspiring politician found himself talking with a young engineer. Naturally the talk turned to the railroad, and the engineer, Grenville Dodge, was full of the subject. So Abraham Lincoln fired questions at him, one after another, and listened carefully to the young fellow's enthusiastic replies. Dodge declared that Council Bluffs was *the* place where the railroad should touch the river. He filled Abraham Lincoln up with facts and figures. Lincoln left Council Bluffs the next day and made the loan. Within a year the votes of free-soilers, Abolitionists, industrialists, and railroad promoters had made him president of the United States.

The Civil War made the completion of the railroad imperative, and in 1863 President Lincoln sent for the young engineer, who was then a general in the Union Army. The president decided that Council Bluffs should be the railhead on the Missouri. The name of the town had been taken from the map of their explorations made by Lewis and Clark.

When Lewis and Clark came up the river, one of their main objects was to inform the Indians that their country had passed from the control of France to the government of the United States. They held council with the Oto and Missouri chiefs on beautiful rising ground covered with short grass on the Nebraska side. The Indians came to visit them August 2, 1804, and fired volleys to empty their guns in token of friendship. The white men fired their cannon, and provided a feast for the tribesmen, to which the Indians contributed "watermillions." Next day the mainsail was stretched for an awning, speeches and gifts were exchanged, and a treaty made. The explorers dubbed the spot Council Bluff. But later, when making their map, for convenience they lettered the name on the east side of the river, so that most people supposed that the council was held on the Iowa side. To this day, it is commonly believed that Lewis and Clark met the Indians on the site of the City of Council Bluffs, Iowa, about twenty miles below.

The railroad brought prosperity to Omaha as well as to

Council Bluffs, and ever since both cities have claimed to be *the* railway center and *the* gateway to the West.

The towns are separated from each other by three miles of yellow bottom land. The poet Wordsworth, in his famous sonnet on Westminster Bridge, writes of the Thames, "The river glideth at his own sweet will." Wordsworth should have seen the Missouri! It glideth back and forth through those bottoms in a most disconcerting fashion, and appears to take a perverse delight in confusing people as to which town is on which bank.

Council Bluffs soon extended into a huge loop of the river. As soon as that part of town was well-established, the river, chuckling mischievously, cut off the neck of the loop, transferring that ward of Council Bluffs to Omaha. The citizens of Omaha promptly moved in and increased the population there—a population which Council Bluffs still regularly adds to its own census returns, contributing not a little to the warm sisterly feeling that exists between the two cities.

Omaha, of course, has the larger population. They are a hardy breed, as people must be to live in a town where many of the streets are as steep as a house roof. If the brakes on your automobile need checking, you can start at the top of the bluff and coast down on a beautiful tree-shaded avenue straight out into the river—if it is still there.

In the beginning, the prosperity of Omaha was due almost entirely to the railroads, which in those days pillaged the country in a manner that made train robbers like Jesse James into national heroes. Mr. George R. Leighton has written an amusing, if somewhat cynical, account of the devices and desires of the men who ruled Omaha, which may interest those who enjoy social satire.[4]

The people of Omaha, however, are not downhearted. They are a tribe of boosters, and their faith is expressed in the elaborate rituals of a unique community organization of civic promoters.

It all began with the State Fair in 1894. Then, night after night, weary women stood about in the streets holding the moist hands of crying children with nowhere to go. Somebody who saw this forcefully pointed out that, unless the city provided evening entertainment for out-of-town patrons, Omaha would never see the State Fair again.

Stunned by this appalling prospect, the merchants formed their booster outfit. It was agreed that whatever was done must be spectacular, big, striking enough to blanket the shopping district—and absolutely *free*. The first plan was for a Harvest Festival topped off by a night parade. Conservative merchants protested, "That will take a lot of cash, publicity, and organization." But the boosters replied, "It will be worth it."

The next question was, "What shall we call the thing?"

Nobody had the answer until one member of the group spoke up: "Why not reverse the name of the state? Nebraska spelled backwards is Ak-Sar-Ben. We are to save the fair for Omaha, so we will be the Knights of Ak-Sar-Ben."

The order devoted itself to winning the good will—and the business—of the people of the state. In those days, the Missouri Valley was a joiners' paradise.

Some four hundred years ago the Spanish conquistador Coronado rode over the plains toward the Missouri, searching for the golden cities of the fabulous Kingdom of Quivera. This fact inspired the scribes who concocted the ritual of Ak-Sar-Ben. Their initiation ceremony is a unique free show combining the horseplay of a college fraternity hell-week and the solemnity of a dignified secret order, all written into the plot of a home-talent comic opera. The novices are assigned unrehearsed parts in the play. These and the spectators are brought by special train from all over the neighboring country on both sides of the river to share in the entertainment, which is held in the order's great "fun plant"—a huge auditorium known as the Madison Square Garden of the Middle West.

The great event is a stunning coronation ceremony at which some civic leader is crowned king of Quivera for the year, and one of the leading debutantes of the season is crowned queen.

Since its organization, Ak-Sar-Ben has sponsored parades, flower shows, harvest festivals, horse racing, and livestock shows, changing its program to suit conditions as the years pass. Its leaders include many of the leading citizens and industrialists of the region, and its slogan is "Knights to the rescue!" Its purpose is to do whatever will make Nebraska "more prosperous, happier, and a better place in which to live." The Livestock Show attracts representatives of nearly a hundred thousand Four-H Club members and has brought into being the superior breeds of hogs and cattle common in all that country.

Omaha is also celebrated for a unique institution— Father Flanagan's Boys' Home—better known as Boys' Town, or the City of Little Men. This is a home for homeless boys, and has educated and trained for useful citizenship thousands of young men who might otherwise have gone to reform schools and into a life of crime. Father Flanagan, an Irish Catholic priest, believes that there is "no such thing as a bad boy."

Fruit trees are so important on the Nebraska bank that there is a veritable tree cult there, of which the temple is Arbor Lodge—a handsome building, memorial to J. Sterling Morton, the founder of National Arbor Day, first proclaimed in 1874 by the governor of the state. Everywhere one finds orchards, canning factories—but even more numerous are golf courses!

Never outside Scotland were there so many fanatical golfers as one finds on the Missouri from Omaha up almost to the Yellowstone. Every city, town, village, and Indian subagency has its golf course. In one such tiny settlement, we used to begin by killing the rattlesnakes on the green before teeing off. Golf and pheasant shooting are the great outdoor sports on this part of the Missouri.

People from other regions have not always understood the origin of breezy Western manners, the love of horseplay and noisy fun characteristic of plainsmen. Yet it is a very old tradition, and, to those who follow it on the Upper Missouri, still seems a valid one.

In old times on the Plains, Indian enemies sneaked up in silence, but Indian friends always came yelling and making a loud noise. That was their custom. After firearms were brought in by traders, Plains Indians naturally made their peaceful salute by firing—and so emptying—their guns whenever they approached a friendly camp or fort or settlement. And their custom was, very naturally—and indeed of necessity—taken up by all traders, trappers, rivermen, and frontiersmen.

This, in fact, is the origin of that old cowboy custom of riding in at a high lope, whooping at the top of the lungs, and "shooting up the town."

So, even today, your truebred plainsman is given to boisterous greetings and silent enmity. Silence on the Plains means dissent, distrust, dislike. And the Westerner's breeziness derives legitimately from that old, old custom of the frontier.

CHAPTER XIII

Chief Blackbird and Sergeant Floyd

IN EARLY times Indians lived on the banks of most American rivers, and so the Indians' history is part of the early history of those rivers, though the redskins who dwelt on their banks have vanished.

The Missouri, however, differs from most other streams in the country in two respects. More Indians lived along its banks in the old days than along most other streams, and they played a conspicuous part in its history. Moreover, on the Missouri, the Indians refused to vanish. They are still there, and still making history—thousands and thousands of them. They are all-out Americans, too; not a rebel or a traitor among them: their sons are among the best fighters in our armed services. On some parts of the river, Indians are far more numerous than white men. Without them, any account of the early days along the Missouri would be meaningless, and they cannot be neglected in the story of the river in our own times.

The traces of their vanished towns and unmarked tombs, emblems of mortal vanities, are found from one end of the stream to the other.

But there is one Indian landmark mentioned or described by all travelers on the river—Blackbird Hill. This grassy hill towers up several hundred feet, overlooking all the countryside, and is visible for miles as it rises above the line of bluffs on the west bank.

The Missouri, washing the base of this promontory,

meandered winding and doubling in a maze of links and loops, leading back almost to its starting place, so that voyageurs in old times, after laboriously navigating for thirty miles under the bluffs with poles or sails or paddles, found themselves at last—as if spellbound—still within a thousand yards of the hill. Perhaps no landmark on all the river remained in view so long. But Blackbird Hill is remembered even more for its romantic story and for the grave that crowns it—one of the curiosities of the Missouri.

Coming up the river from Council Bluffs, high land appears on the west bank—bluffs of yellow sandstone known as the Yellow Banks, which continue for miles along the stream. These bluffs, now called the Blackbird Range, were haunted, as Audubon reports, by many bank swallows.

From afar, boatmen could see on the top of the hill the mound of this remarkable tomb, from which a bare pole leaned forlornly. Most men on their first trip upriver climbed the hill to inspect the tomb of the chief for which it was named. The hill rose above rounded elevations and bluffs of various heights. Visitors were struck by the wide view of the broad river valley with its wooded islands and the prairies across the stream.

On the topmost knoll was the grave—a mound some twelve feet in diameter at the base and six feet high. On the eight-foot pole stuck in the top, Lewis and Clark fastened a white flag bordered with red, blue, and white, as a tribute to the famous chief of the Omaha tribe who was buried there.

Blackbird was captured by the Sioux in his youth. Afterward, returning to his own people, the Omahas, he became a warrior of note. He had an insane vanity and love of public admiration which made him utterly fearless in battle, so that he dazzled his tribesmen by his daring exploits.

But Blackbird owed his power and fame to the traders. One day an old man was taken down to St. Louis. He came back declaring that he had been made a "great chief" by the white men, and immediately began to name "soldiers" to help him control the Indian trade of the tribe. Blackbird, ambitious and promising, was chosen by the chief as one of these soldiers, and accompanied him on a second visit to St. Louis.

Blackbird was a handsome young buck, and caught the fancy of the white men. On sounding him out, the traders found he would readily become their tool for the exploitation of his people. Forthwith they gave him a medal and a paper commission as "chief" of the Omahas. Blackbird returned, announced his new authority, and immediately showed himself smarter—and more unscrupulous—than all other "trade chiefs" appointed by the whites.

When the first trader showed up, Blackbird was on hand at the landing and told the trader to bring all the packs to his lodge.

When this had been done, Blackbird opened all the packs and took for himself whatever pleased him—nearly half the goods. The trader began to look glum, seeing all his profits stolen in this highhanded way.

But Blackbird reassured him. "My son," he said, "what I have taken are my goods; what you have left are yours. Do not cry. I will make my people trade with you on your own terms."

Blackbird then ordered his crier to climb to the top of the earth lodge and shout to the village that every man must immediately bring all his furs and skins to Blackbird's lodge to trade, warning them that they must take what the trader offered and not haggle.

When the trading was done, Blackbird had fully earned his huge commission. The trader rubbed his hands and afterward said that this was the most profitable trip he had ever made. Naturally, after that, the traders favored the "chief."

Of course Blackbird could not impose on his people

like that with impunity. So, when they grumbled and threatened, he turned to the traders for backing. Now, it was never the wish of the traders to cause open strife within a tribe and so split it, if they could trade with it as a unit. So, among them, they found a way out of this difficulty. One of the traders gave Blackbird a supply of arsenic, and told him how to administer the poison.

From that day Blackbird began to acquire a reputation as a medicine man gifted with supernatural powers—and particularly the power of prophecy. Whenever one of the warriors opposed him or objected to his arbitrary deeds, Blackbird solemnly prophesied that the offender would die within a certain time—and every time his prophecy came true. He soon became such a terror to his people that they dared not offend him in the slightest way; when it was necessary to wake him from sleep, nobody ventured to shake him or call his name. They would tickle his nose with a blade of grass.

Little Bow, well-known and well-liked in the tribe, was the only man who dared oppose Blackbird. The people trusted him, and when he opposed Blackbird, it was a serious challenge to the chief's reign of terror. Blackbird saw that he must make away with Little Bow.

Long tells the story: When Little Bow returned home after a hunting trip, and his wife was preparing dinner, he noticed something peculiar in her behavior which made him suspect something was up. When she put the bowl of food before him, he did not eat, but questioned her. Alarmed by her evasions, he forced her to eat the meat herself. Dying, she confessed that Blackbird had frightened her into mixing a dose of his terrible medicine with the food in order to kill her husband.

The secret of Blackbird's power was now uncovered. Little Bow rounded up two hundred of his people, who moved away and built a separate village of their own elsewhere on the river. The tribe was split in two until Blackbird's death.

As time passed, he became more and more tyrannical. Finally he murdered his own wife in a fit of anger, but was never punished. He got very fat in his last years. But such was his authority that when he went to a feast, four warriors carried him, each holding one corner of the buffalo robe on which he lay.

In 1800 smallpox struck the Omaha camps. Two-thirds of the tribe died of the plague or in their despair committed suicide. They even set fire to their village. Blackbird was stricken with the rest.

When he saw that he was dying, he gave orders for his burial.

With Indians, death is the greatest event in life—the one strictly personal experience—and great men among them generally spared no pains to make a good exit. Accordingly, Blackbird called his surviving relatives around him, had himself dressed in his finest battle dress of fringed and beaded buckskins and crowned with a superb war bonnet. He had his face painted. His best horse was brought out and saddled, and as soon as the chief had died, they tied him in the saddle upright, fastened his bow in one hand, hung his shield and quiver over his shoulder, with his pipe and his medicine bag, with a supply of pemmican and tobacco, his flint and steel and tinder, the trophies and scalps he had taken, and so carried him to the top of the hill. There a shallow cave or hole was found or dug, and the horse led into it, facing so that, as Blackbird had requested, he "might still see the white men coming up the river to trade with his people."

Then stones and sod were laid up around and over the rigid horseman and his mount.

The story goes that the horse—or mule—was buried alive, but this seems unlikely, if Blackbird hoped to ride its spirit into the happy hunting grounds. It is more likely that the animal was strangled in accordance with Omaha custom after the mound had been sufficiently raised to support its body. Ordinarily when horses were killed at a grave, they were not buried with the owner. But Blackbird was an original man and vainglorious, and he may have dreamed that he should be buried so. If so, the dream would have been sufficient authority for this departure from custom. Indians obeyed their dreams. Perhaps the chief merely wished to outdo all others.

This is the story of the burial of Washinga Sahba, who died only four years before Lewis and Clark visited his tomb. The sentimental Catlin was indignant at the tale, accusing his good friends, the traders, of conniving at the poisoning of their customers. When Catlin visited the tomb in 1832, he penned a strong defense of the "noble chieftain"—after digging into the grave and stealing his skull! They say the cranium is now in the Smithsonian Institution at Washington. And so Blackbird is now but pyramidally extant. There is such a thing as being buried too conspicuously.

But that old scoundrel would not care what became of his bones, so long as he was remembered. His hill is on the old Omaha Reservation near Macy, Nebraska, some seventy-odd miles upriver from the city of Omaha. His spirit must be delighted at the number of white men who flock to visit it.

Catlin was not the only painter who made that hill into a picture; almost every artist who went up the river in the old days stopped to sketch it. The best of these was the landscapist, Karl Bodmer, who accompanied the expedition of Prince Maximilian. Better than any of the early painters, Bodmer caught the scale and vital grandeur of the Upper River.[1] He had known the Rhine.

The two most famous tombs on the Missouri River are found within a few miles of each other on this part of the stream. The second is almost within sight of Blackbird's, farther up, on the east bank. This is the grave of Sergeant Charles Floyd, a trusted member of the expedition sent up the stream by Thomas Jefferson and led by Lewis and Clark.

The explorers had passed Blackbird Hill and the abandoned Omaha village on their way up the river. Anxious to make a treaty, and not finding any natives about, they set fire to the prairie. This, the customary signal of a trader seeking customers, brought no Omaha; but a number of Oto and Missouri Indians turned up and were made welcome. The Indians proved friendly, and all was peace and joy.

The white men were happy also. They had seined a creek nearby, and had made a great haul. The Missouri Valley is a great breeder of fish; no state in the Union produces more freshwater fish than Missouri. Sergeant Floyd and his comrades had caught upward of eleven hundred: pike, bass, trout, red horse, buffalo, rockfish, catfish, perch, flatback, silverfish, and something resembling a salmon—not to mention shrimp and any amount of good fat mussels. They were feasting.

It was a hot August evening. But the Indians, anticipating the bounty of their hosts, celebrated with a lively dance on the sandbar. Some of the white men, in equally high spirits, joined in. No doubt each party tried to outdo the other.

Sergeant Floyd, one of the most vigorous of the dancers, finally became overheated and stretched out on the cool sand to rest.

Next day he was suddenly stricken with illness—diagnosed as "a bilious colic." He grew rapidly worse, in spite of all that officers and men could do. All that night he suffered.

Next morning the Indians, happy in the possession of a canister of whisky, left them. It was a fine day, a southeast wind was blowing. They carried the sick man aboard and set sail. After passing two islands, and making thirteen miles, they lay to under some bluffs on the left bank.

Here Floyd died. We may quote from Clark's original journal entry:

"Serjeant Floyd is taken verry bad all at once with a Biliose Chorlick we attempt to reliev him without success as yet, he gets worse and we are much allarmed at his situation, all attention to him . . . Sergeant Floyd much weaker and no better .. . . as bad as he can be no pulse and nothing will stay a moment on his stomach or bowels. . . . Died with a great deel of composure, before his death he said to me 'I am going away I want you to write me a letter'—We buried him on the top of the bluff ½ mile below a small river to which he gave his name, he was buried with the Honors of War much lamented, a seeder post with the Name Sergt. C. Floyd died here 20th of August 1804 was fixed at the head of his grave—This man at all times gave us proofs of his firmness and Determined resolution to doe service to his countrey and honor to himself after paying all the honor to our Deceased brother we camped in the mouth of *floyds* river about 30 yards wide, a butifull evening." [2]

Floyd was the only man who died on this expedition of more than two years; he was also the first United States soldier to die within the boundaries of the Louisiana Purchase west of the Mississippi.

The bluff on which he was buried was afterward known as Floyd's Bluff.

The account of his death, in spite of Clark's peculiar orthography, became widely known. And though the cedar post set up to mark his grave was burned away by prairie fires, it was regularly replaced by new markers as regularly whittled away by souvenir hunters.

The marker could be seen by boatmen from a considerable distance, and the beautiful site, covered with grass and wild flowers, often attracted visitors.

By 1856 the river had undermined and cut away Floyd's Bluff until the next year's freshet wiped out the last hundred feet between the river and the grave. The skeleton lay exposed somewhat more than one hundred feet above the water.

Citizens of the nearest town made haste to rescue the remains. A man was let down on a rope over the brink of the bluff to attach a cable to the box. The bones were then hoisted up and reinterred on May 28th with appropriate ceremonies at a spot about two hundred yards farther from the river. A new coffin was made of black walnut timber growing nearby.

The Missouri has continued to eat away the bluff until now the original site of the tomb is out in the air above the surface of the water.

On August 20, 1895, on the ninety-first anniversary of Floyd's death, a monument was dedicated, in the form of an obelisk fully one hundred feet high. This taper shaft can be seen for miles up and down the river. Because of its beautiful site and wide prospect, the spot has become one of the show places of Sioux City. It stands in one of the parks there.

Captain Meriwether Lewis, in reporting to the War Department when the expedition was over, called Floyd "a young man of much merit" and recommended that his father should receive a gratuity. Floyd's journal was long lost, but afterward discovered in the archives of the Wisconsin Historical Society, in a collection of documents from Kentucky, Sergeant Floyd's home state.

Many another grave besides Floyd's has been destroyed by the Missouri River: graves of prehistoric people, graves of Indians, graves of the early pioneers, graves of more recent settlers. These, unknown, unmarked, had no friends to rescue their tenants and bury them on safer ground; whole cemeteries have been undermined and swept away by the ghoulish waters of the stream. The Missouri is a regular body snatcher.

A good many of the people who went up the river had a similar lust for grisly trophies; records of old times frequently mention persons collecting Indian skulls, either for themselves or as gifts for friends at home. Larpenteur mentions a "requisition for Indian skulls ... made by some physicians from St. Louis."

Not long before, an Indian who had boasted that he was bulletproof was shot by a brave who wished to test the power of his medicine. Riddled, he was buried near Fort Union. This was the Indian painted and described so vividly by Catlin under the name Wi-Jun-Jon (the Pigeon's Egg Head). Larpenteur reports: "His head was cut off and sent down in a sack with many others."

Audubon tells how he and the fur trader, Mr. Denig, "walked off with a bag and instruments, to take off the head of a three-years-dead Indian chief, called the White Cow." They dragged the body from the tree in which it had been placed, took off the head, and left the desecrated remains lying on the ground. Audubon coolly remarks of the headless chief, "He was a good friend of the whites." [3]

Apparently, collectors of such trophies preferred skulls of Indians whose name and tribe were known.

But army surgeons were among the most inveterate head-hunters on the Missouri. Tales are told of how certain of these gentlemen cut the heads off Indians killed in battle and boiled them in the camp kettle until the flesh was gone—"for scientific purposes." One surgeon was known among his fellows as "the head-boiler."

Since few Indian bodies were buried underground, it was a simple matter to find and secure a skull. Such was the race prejudice of those days that, though the dead man might have been well-known to those who mutilated his body, there was none to say "Alas, poor Yorick."

CHAPTER XIV

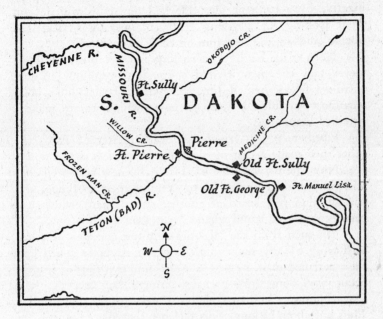

The Petrified Man

ALL SORTS of bizarre things happen on this wild river, which—once you stop to look at it—turns out to be as strange as the Congo or the Zambesi. In the early nineties, when the whole country was talking about the World's Columbian Exposition to be held at Chicago, it occurred to some gentlemen on the river that there might be fun—and money too—in an exhibit from the Missouri. The problem was to find something to exhibit.

One day in Forest City the butcher, one Bill Sutton,

slicing away at the beef on his block, was suddenly struck inactive by a brilliant idea. He laid down his whittle and went looking for a young physician, seeking advice. The doctor laughed and sent him on to William Horn.

Horn made his living as a lime burner, and after some secret talk the men invited a fourth conspirator to join them—a fellow named Jim. They decided that Jim's body was most suitable for their purpose—which was to make a petrified man.

Obediently Jim lay down. With plaster of Paris, the other men made a cast of his body.

Now the perpetrators of this hoax knew very well that scientists would be called in to examine their specimen, and perhaps they had heard of another stone "man" which had been exposed as a fraud when sawed into.

The men from the Missouri were not going to be taken into camp by any such simple trick. If anybody sawed into their petrified man, he was going to find evidence of human occupancy. Somehow—perhaps through the doctor's aid—they obtained a human skeleton. It was a little smaller than Jim's but about as long. This skeleton they placed within the cast, which they then filled up with cement.

In great secrecy the trio hauled the heavy image out to the bank of the stream near Forest City, where limestone abounded.

There, some days later, while out looking for lime to burn, Horn unearthed the figure and immediately announced his amazing "discovery."

That discovery was given wide attention in the press, and when the Forest City man arrived in Chicago, thousands flocked to see him. Afterward he was carted around the country and exhibited—a solid Western practical joke on the folks back east.

Nature herself hoaxed scientists on the Missouri. On Bon Homme Island and on the shore adjacent, Nature laid out earthworks of sand so regular that the explorers Lewis and Clark carefully mapped them as ancient "fortifications" supposed to have been built long before. They may have been built long before, but they were not made with hands. In those days few Americans had any notion what the wind in the Missouri Valley could do. We, in our time, have unhappily a better understanding.

Queer things happened around the "Fork"—Where the Cheyenne River came in. Not far up the stream on Cherry Creek, a young Sioux matron had a violent quarrel with her man. It was wintertime, cold and snowing, but she was so angry that she left the camp, taking no food and only the clothes she wore, striking out blindly into the storm. All that day and night she kept on, hungry and cold, trying to get as far away as possible before the snow stopped falling; for then, she knew, her tracks would be visible and her relatives would be able to find her. She was bitterly determined never to go home again.

When the snow stopped falling, she halted, fearing to make tracks. Then she noticed that the wind had swept the slope of the hill above her partly bare. She thought she would go up there, away from the trail, where she could leave no tracks.

As she went on, tired, chilled through, and hungry, she stumbled on the mouth of a small cave. It appeared to be empty. She crawled in, pulled her blanket over her head, and fell into a shivering sleep.

After a bit she ceased to shiver, dimly feeling a glow of warmth on either side. Warm fur touched her cold hands. She relaxed and fell sound asleep.

When she woke in that narrow den, she looked up at the yellow eyes, white fangs, and lolling tongues of four big gray buffalo wolves. They sat on their haunches close about her, regarding her with fixed attention. Frightened, she caught her breath and sat up. The wolves got to their feet, seemed to smile, and whined a little, making friendly noises which she interpreted, in her Indian way, as human talk. Her fear melted. She spoke to them.

The wolves left her to hunt, and when they returned, the biggest came dragging in a chunk of raw meat. She was too hungry to mind that—she ate her fill. *Washtay!* That was good.

Having been thus accepted by the wolves, she set up housekeeping among them and every day prepared the meat which they brought in, making pemmican and clothing for herself from the hides of deer pulled down by her new furry brothers.

Indians on Cherry Creek who remember her say that she stayed one whole winter in the wolf den, but that no Sioux ever discovered her hideout. Chief White Bull, who knew her well, told me that she cooked her own meat, building a fire in a hole at night so that no one could see her smoke. Being a well-trained, competent young woman, she managed to live quite comfortably, and quite as happily as ever did Mowgli or Romulus.

All things come to an end in time. And one day it seemed to her that the wolves told her she must go. She was not to live with them any more.

But Woman-Who-Lived-With-Wolves was in no mood to return to her family in the Sioux camp. She no longer felt any love for her relatives, and she knew that her marvelous experience had set her off from all other women in the tribe. She was now a medicine woman. Still, she was lonely without the wolves.

One spring morning a party of young men from the Sioux camp rode out looking for wild horses. They saw a cloud on the prairie—so much dust that it was hard to tell just what was causing it. But as they slipped up the coulee toward the dust, they saw it was a band of mares closely guarded by a handsome spotted stallion. The young men uncoiled their lariats and quirted their ponies into a run. They hoped that they might capture some good horses, for the grass was not yet high and the animals were lean and might tire quickly. Charging in from three sides, the hunters gradually overtook the herd. The frightened mares would have scattered, but the stallion, racing from side to side of his galloping harem, nipped and pawed at every straggler and kept the mares in a compact bunch. He set a fast pace.

Finally, however, the Sioux closed in and separated the stallion from his mares. Then all the animals scattered over the prairie.

Soon after, one of the hunters dimly saw through the dust cloud some creature—not a horse—running among the animals. He could not tell what it was at first. Then he saw it was an Indian woman.

He called to her. But she did not stop; she ran on and on, sprinting over the prairie. All the riders took after her. One of them dropped a noose over her shoulders and stopped her in her tracks. Others rode up, jumped from their horses, caught her arms. She fought like a wild thing, snarling, biting, clawing, glaring wildly at them through her long flying hair; but she had no chance. They held her. They recognized her as the runaway wife.

Some of the young men thought she must have been running with the horses. But White Bull told me this seemed unlikely. Said he, "How could a person get meat while running with a herd of horses? She would starve—even if the animals had been willing to let her run with them."

It seems more likely that she just happened to be in the path of the wild horses as they fled from the hunters, and that they passed and left her in the dust to be taken by the young men.

The hunters took the wild woman back to camp.

There her relatives made her welcome, but she, wary as a wolf, sat still, making no sound. But the women of her family gentled her, combing her hair, painting her face, dressing her in new clothing, and so by slow degrees restored her to something like a normal state of mind.

At length she told her curious, pitiful relatives something of her strange adventures. Other people dared not question her. She was now a famous medicine woman, a person of supernatural powers—powers that were feared, envied, and sought after. For to the Indian, anything sufficiently unusual to seem unnatural was believed supernatural. Apparently she used her powers to help her people.

Other persons in the tribe who had been favored with gifts of mysterious power naturally wished to test the strength of this new rival, and a contest was arranged at which White Bull and Boat-Upper-Lip challenged Woman-Who-Lived-With-Wolves to compete with them.

This contest was held in public, and each contestant exerted his power to "kill" or overcome the power of his opponents. They stood some distance apart, and, holding missiles charged with power in their palms, "shot" these at their rivals to see whose power was strongest. Boat-Upper-Lip and Woman-Who-Lived-With-Wolves combined to overthrow White Bull. They shot bumblebees, grasshoppers, and rolled up buffalo hairs at him. But his power was strong and he withstood them, keeping his feet and his courage.

When his turn came, White Bull faced his two rivals. He was encouraged by the fact that they had not been able to make him pass out; but it remained to be seen whether his medicine was strong enough to bowl them over.

Staring fixedly at them, he held in his hand some of those small worms which come out of the forehead of an elk. Gesturing at them as though throwing these, he "hit" them both. They could not repel his medicine. First one and then the other fell unconscious. Victor, White Bull then magnanimously used his power to revive his defeated rivals. He drew the missile worms out of their bodies and brought them to. After that everyone in camp believed that the elk was a real helper to White Bull—his medicine was strong.[1]

In spite of such a public setback, Woman-Who-Lived-With-Wolves was highly respected throughout the rest of her long life. Some of her relatives still live along the Missouri, and your Indian friends will, if you like, lead you to the cave where she kept house through that long winter for her brothers, the wolves.

Of course stories are not uncommon among Indians telling how wolves and other animals have befriended human beings, guiding them to game or to camp, giving warning of enemies by howling in a certain way, and the like. But the greatest Indian marvel on the river was Calf-Woman. This strange being was born in the winter of 1849-50 on Cheyenne River, somewhat above the mouth. The event is pictured on nearly all old Sioux winter counts or calendars painted upon buffalo robes.

That day a party of Minniconjou Sioux hunters were running buffalo. One of them shot a cow, and his companion dismounted to help butcher the animal, and so earn a share of the meat. There was snow on the ground and the men were cold and hungry. The two of them went busily to work, stripping off the skin and cutting the carcass to pieces. When the belly was opened, they found there the bag which held the unborn calf in it. Such a tender calf was a favorite dish among the Indians of those days.

One hunter went on butchering the carcass; the other eagerly ripped open the bag. Then, amazed and terrified, he jumped back. Instead of a calf, an old woman with human features and long gray hair sat up and stared at him for a long moment. Then she rolled over dead.

When the two shaken hunters had collected their wits and found their voices again, they began to yell and shoot, beckoning to their fellows to come and see that marvel. The whole party hurried up and dismounted to examine the extraordinary creature which—or who—was so altogether like a human being.

That uncanny thing made the men stand silent, trembling, with their hands over their open mouths. At last an old man ventured to advise them. The two hunters plucked some sprigs of sage, a sacred plant, and wiped themselves all over—a rite of purification. They did not touch the meat of the cow or Calf-Woman again, but hurried away leaving them lying on the snow. That day there was no more hunting along the river. . . .

Higher up, the Missouri has one of the most remarkable battle monuments in the world. In fact, it is unique.

One day the Sioux in great force came riding over the prairie to attack a town of sedentary Indians on a bluff above the Missouri. It was a surprise attack, for the townsmen had not expected that the Sioux would be raiding in such rainy, muddy weather; one of the men was out alone on the prairie looking for his pony. Yet here they came, yelling and shooting, on a dead run over the wet prairie, trying to catch him. He took one look and sprinted for home.

The warriors in the town, outnumbered and unprepared for such a sudden attack, could not rally in time to save their desperate comrade.

But one warrior, more daring and quicker than the rest, snatched up his gun, leaped on his horse, and sped out of the palisade. He circled across the wet grass toward the enemy, trying to cut between his desperate friend and the horde of charging Sioux. As he raced on, the leading Sioux rode to meet him. Both fired together. The Sioux fell dead, plop in the mud; his horse galloped away. At that, all the other Sioux reined up. The victor offered his hand to the breathless runner, who leaped up behind his rescuer. Both reached the village safely. The Sioux packed their dead chief on his pony and rode away.

Next day, after it was certain that the Sioux had gone, the women of the village went out onto the prairie and, finding the tracks of the horses and the men and the mark of the dead Sioux outlined in the mud where he had fallen, they took their hoes and went to work to deepen and enlarge every mark or bit of sign which showed the progress of the fight. The whole battle was thus indelibly stamped on the prairie, and after nearly a hundred years the story could be read there. Such a monument, unless plowed under or washed into the river, might well outlast many a more pretentious memorial. . . .

Of course, there are also many mythical adventures associated with the Missouri; some of them ancient, some of them almost as new as yesterday's paper. Blackfeet in Montana have hundreds of tales about their famous small-time god and trickster, Old Man. The Sioux tell similar stories of Iktomi, a name also applied to the spider because he is so skillful and lives by his wits, managing to kill game without leaving his lodge—as all Indians vainly dreamed of doing. Other tribes celebrated similar exploits of Coyote. There are literally thousands of these yarns, and the hero, who is usually also the goat, behaves much like Donald Duck, whose tricks so often recoil upon him. Sometimes, however, the trickster wins out; he may even prove helpful sometimes, after the manner of Mickey Mouse. The Indians are therefore quite ready and able to appreciate such movie stars. In fact, a new baby in a celebrated Sioux family I know was recently christened Mickey Mouse!

White men on the Missouri have mythical heroes too: Barada of Nebraska and his rival, Febold Feboldson, who bid fair to outdo Paul Bunyan. There is a world of folklore, so that almost every butte and tributary along the stream is connected in the minds of those nearby with some ghost or hero, some hanging or rescue, some jest or anecdote, some tragedy or portent. To record them all would require volumes. Hundreds have already been recorded.

Curious institutions flourished on this river. One of these was the Warrior Society. Lewis and Clark first heard of it among the Yankton Sioux: "an institution peculiar to them, and to the Kite [Crow] Indians ... from whom it is said to have been copied ... an association of the most active and brave young men, who are bound to each other by attachment, secured by a vow, never to retreat before any danger, or give way to their enemies. In war they go forward without sheltering themselves behind trees, or aiding their natural valor by any artifice. Their punctilious determination not to be turned from their course, became heroic, or ridiculous, a short time since, when the Yanktons were crossing the Missouri on the ice. A hole lay immediately in their course, which might easily have been avoided by going around. This the foremost of the band disdained to do, but went straight forward and was lost. The others would have followed his example, but were forcibly prevented by the rest of the tribe."

The Upper Missouri was also the stream on which we find a vessel unique in North America. It was seldom seen on any other stream, and then only in the hands of persons who had known it on the Missouri. It is the American version of the primitive Old World coracle. Here it was called bullboat.

The bullboat was simply an open-work basket made of willow sticks lashed together with strips of green rawhide and covered with a single buffalo skin.

To manufacture it all that was necessary was a freshly killed buffalo and a thicket of willow shoots—and, of course, a paddle. To make a bullboat the voyageur cut a dozen or more straight willow shoots almost as thick as his wrist. He laid five of these, each about ten feet long, side by side eight or ten inches apart. He then placed four others similarly spaced at right angles to the first five and tied the sticks together where they crossed with strips of green hide. The free ends of all the sticks were then bent upward and lashed to a circular or elliptical hoop or gunwale in the same manner. Another hoop was bound around the bottom and sometimes one or more hoops at intervals between the bottom and the gunwale. That done, he turned the "basket" upside down, stretched a raw buffalo hide tightly over it, and lashed the edges securely to the gunwale. Bull hides were preferred because they were thicker and tougher than cow-hides. The hairy side was the outside.

A few hours in that dry climate shrank the rawhide lashings and covering of the boat so that all was tight and stiff as a board. The outside might then be smeared with melted tallow to prevent softening by the water. When he had finished, the voyageur had a leather tub two or three feet deep and four or five feet in diameter, depending on the size of the hide.

The whole contrivance weighed only twenty or thirty pounds and was easily carried on a squaw's back or in the basket of a pole drag or travois. On reaching the bank of the river, the boatman dropped the coracle into the water, stepped carefully into the middle of it, and knelt down on his duffel at the offshore side of the vessel. Leaning forward, he thrust his paddle into the water ahead and drew the blade back toward the boat or paddled first on one side and then on the other toward the rear. In this way the bobbing craft was made to advance, swinging a little first to one side and then to the other according to the pull of the paddle. Such a boat would carry several people or one person and a load of one or two hundred pounds safely across the stream. Men—and women too—did not fear to cross the Missouri in such boats even at high water or when the wind had raised considerable waves. Travelers have described how, from a little distance, bullboats would disappear behind the waves so that only the squaws who paddled them were visible above the water, appearing like so many mermaids.

Such boats were commonly used around Indian villages and trading posts and occasionally for long trips downriver. Their shape made it difficult to paddle them against a strong current, but what they lacked in gracefulness and speed they made up in capacity and ease of manufacture. Usually the bullboat was not perfectly circular and might be a foot or more longer than it was wide, conforming to the natural proportions of the buffalo's hide. The bullboat traveled better with a load large enough to set it fairly deep in the water. When empty it bobbed around at the mercy of wind and current.

Often the tail of the buffalo (if still attached to the hide) was left projecting from the gunwale of the bullboat to serve as a sort of handle in launching or carrying the craft. Bullboats could be carried upside down, resting on the head, or slung on the back right-side-up with the bottom across the squaw's shoulders, all supported on a strap across her chest. Paddles were much like those used in canoes today with a knob at the end of the handle to fit the hand and a rounded blade, often painted or incised with the owner's mark.

A hundred years ago, travelers used to go upriver to the Mandan town to see the White Savages, supposed to be descended from the Lost Tribes of Israel. In those days there were many children on the Upper Missouri with gray eyes and light hair. I asked Uncle Billy, an old riverman, what he knew about that.

That old water dog tilted his head and shot tobacco juice into the brown water sliding by. "Nowadays," he said, "if you're lookin' for white savages, you go downriver, not up. Tie up at any tall town on the river and take a squint along Main Street. You'll see 'em—plenty. White women, and all gone native, outsquawin' the squaws. Short skirts, bright colors, cheap jewelry, loose hair, plucked eyebrows, bare legs—and paint! Why, some of 'em use more paint than would do an Injun buck all winter. They ride 'straddle, they gamble and drink and smoke, and make faces into little lookin'-glasses all day long, and the way they dance would make Settin' Bull turn in his union card. Happen Lewis or Clark was to turn up on the Lower River tomorrow, he'd swear he was back in the Mandan village."

"Maybe the Indian women had something, at that," I ventured.

Suddenly Uncle Billy turned an animated eye on me. "Maybe you're goin' downriver?" he asked. "Come to think of it, I ain't been to town in a coon's age. Have you got room for me?"

"Plenty," I replied. "But what do you want to go to town for?"

Uncle Billy shot out his wrinkled neck like an old mud turtle. He grinned. "Well, I just took a notion. Maybe I could see some of them White Savages!"

CHAPTER XV

Range and Grange

WHEN the English made their first permanent settlements on our Atlantic seaboard, a curious thing happened. Apparently those early settlers, traveling light, found the whole of English civilization too heavy to carry so far across the sea. So they cut the creature squarely in two, and each group brought over its own half. The Virginians brought the country estate, the landlord, the politician, the lawyer, and the tenant; the New Englanders brought the village, the schoolmaster, the divine, the yeoman, and the merchant. It is a wonderful proof of the vitality of British culture that both halves survived this vivisection and throve mightily. They still thrive, so that each part remains convinced that its own half is the head, and the other half the tail. They could not kill the snake, but they scotched it so deeply that it will be long before she'll close and be herself.

As these two ways of life extended themselves westward, though tenuous and diluted, they overwhelmed all barriers—until they reached the Great Plains. There both were halted abruptly by a country, a climate, a tradition, and a material culture which could not and would not be assimilated. On those vast and perilous plains all men, rich and poor alike, had to meet the same strict conditions; the great man was great simply because he did what others did, and did it better—like Achilles, like Ulysses. Those Greek heroes would have felt perfectly at home with Sitting Bull or Buffalo Bill or "Dad" Lemmon. But can anyone imagine

George Washington or Benjamin Franklin at ease on the Great Plains? The mere suggestion is fantastic.

Nearly all the institutions of Virginia, of New England, were at first transplanted bodily to the plains and withered and died there. At first they lived for a while on the mere sudden wealth of a newly opened country. But as the plains economy became stabilized and the margin of that new wealth was spent, these imported institutions began to wilt—having no roots in that country. Those which remain have been kept alive by subsidies from outside.

Outsiders feel sorry for the Plains as a region of sparse population, meager natural resources, limited political influence, and economic dependency. It is perfectly true that the population is scattered over great distances. There is one stretch of two hundred miles along the Upper Missouri where there is scarce a village worthy of the name. It is true, too, that on the plains the natural resources are few and scattered. The people, being so few, cast few votes. And since whatever they buy or sell is arbitrarily priced in remote markets, they are economically dependent. And so outsiders feel sorry for the plainsmen.

One might as well feel sorry for a steer because he has no britches. He was never intended to wear britches, any more than the Plains were intended to support Southern plantations, Midwestern farms, or Eastern industrial towns. The Negroes from the South knew this before the financiers did. The Negroes came to the east bank of the Missouri and stopped; they knew that was the end of their country. The small farmer came into the plains, broke his back and his wife's heart, and trekked back to where farms belong. The labor organizer and industrialist crossed the Missouri, but found little welcome. What cowboy ever wished to join a union?

The institutions of the South and the East and the Middle West do not belong on the Plains. There a new pattern was imperatively presented to the settler. The blueprint for Plains civilization is drawn so sharply and definitely that men cannot vary from it and survive. Yet this fact—that the institutions successful elsewhere in the United States do not thrive on the Plains—is no sign that the Plains are barren, hopeless, without culture or tradition. On the contrary, the plainsmen have all that is needed for the good life —if they could only get from under the cumbersome and costly institutions blindly thrust upon them by the rest of the country.

After all, the test of the pudding is the eating. The cowboys and Indians who people the Plains do not migrate. Elsewhere, farmers and factory hands are continually on the move. Happy men imagine heaven in terms of the life they lead—as cowboy songs and Indian stories abundantly prove. Dissatisfied men always think of their way of life in other terms. We have only to look at the novels and stories about factories and farms to discover how life off the Plains is regarded by those who attempt to interpret it. From Hamlin Garland and Upton Sinclair onward, such literature represents the life of the farmer and the laborer as utter hell. No doubt this is a partial picture, but there is a moral in it.

The South dreams of the past, the East sees visions of the future, but the plainsman is happy here and now. He is not very well-to-do. Sometimes his life is one of extreme difficulty. Yet all he asks is a little *more of the same*—more rain, more grass, more cattle—*not* another kind of life. How many farmers or welders or bond salesmen look forward to going on with their present work for all eternity?

You would probably be safe in offering a dollar for every despondent cowboy who can be found, provided someone would pay you a dime for every happy one. When a cowboy has cash, he spends it for a new saddle or a new pair of boots—the tools and livery of his profession. And because the cowboy is pleased with his life and his work, other people take him at his own valuation. He is accepted, all over the globe, as the real American, the national hero. Uncle Sam may still stand vaguely for the U.S.A., for old time's sake; but when people at home or abroad *dream* of America, they see the image of the cowboy.

Millions of words are published every year about him; miles of film are devoted to his mythical exploits; hundreds of people wear his clothes and join in his activities as well as they are able on dude ranches; millions of people flock to watch his rodeos; pictures and statues of this democratic man on horseback can be found everywhere. Half the magazines in the country and half the motion-picture producers would have to go out of business if it were not for him. As a representative of North America, he is out in front and the rest are nowhere.

When Europeans come to our shores, they ask to see cowboys; German prisoners of war demand western stories and books. To make the verdict in his favor unanimous, the cowboy himself spends his Sundays on the ranch reading cowboy stories. There may be no "great" individual cowboy; but as a type he leads the field. This is the universal opinion of mankind.

The region and tradition which produce such a dominant type are not negligible; there is more to the plainsman than meets the eye. The tradition is old and fine in itself, it ripened long ago, and requires no dressing up with borrowed "culture" to gild the tumbleweed. And since the region still compels strict adherence to the conditions which produced that tradition and that type, both remain clear-cut, however great the changes in the world outside. People "go for" a man who, like Santa Claus, is always the same.

His tradition is one of freedom, of loyalty almost feudal to his boss, of courage, fun, and respect for decent women. He feels that he is different from, and superior to, most men, and emphasizes his status by a lingo and a humor all his own.

Hospitable as an Indian, he has the Indian's pride and independence; he covers the ground he stands on and does not permit anybody to shove him around. And when he tells a tall tale, he does not magnify the boss into a Paul Bunyan, like some wage slave, but makes *himself* the hero of the yarn. He knows that courage is freedom; freedom, happiness. Nobody, probably, in North America so fully embodies the old American independence and the old American romantic faith in the natural rights of man.

South Dakota is the "most Republican" state in the Union; North Dakota is the state where the Non-Partisan League arose. But the cowboy made friends on one side of the river—the west side—and is much the same everywhere on the Plains. Politics are for townsmen and farmers, not for the saddle-born. That is true from the Mountains, from Three Forks, down to Nebraska and Kansas.

A good many cowboys, seeing how the country changed, have agreed with Charlie Russell:[1] "I wish the coyotes would get big enough to eat the damn farmers!"

Above Nebraska the course of the Missouri River swings toward the west, then north, cutting South and North Dakota squarely in two, separating the farms from the ranches by a natural boundary. The bottom lands along the river, sheltered under the bluffs, are rich and productive. But once one gets onto the high prairie and moves eastward, one finds a good deal of unplowed land, some of it ungrazed. Coulees and "draws" cut through the rolling or broken semi-arid country, where farms are large though crops are small. One has to keep going east a good way through the wheatlands and small-grain farms to reach a truly agricultural land like that of Iowa or Minnesota (with windmills, silos, big red barns, and dairy herds) checkering the earth.

Along the east bank of the Missouri agriculture must be diversified. The soil is too dry, the rainfall too uncertain, and the hazards of the farmer too numerous. Severe droughts recur with maddening regularity about every twenty years, and so this whole country has been populated and repopulated over and over even during the little more than fifty years it has been under the plow. The disasters of 1894, 1911, and 1933-1936 not only drove out the farm population from large areas, but blew away the very soil. On the other hand, there are good years when rainfall brings bumper crops. There was a good one in 1927. The farmer along the river in the Dakotas need not stretch his imagination to envision Pharaoh's seven lean years.

Every day in such years the sun shines on the parched land, burning the rich soil to a fine grit, sometimes cracking it so that a man can stick his fist between the barren plates. That soil is a gray gumbo, and, when wet, incredibly deep and sticky. A stray horse cannot wander far because in half an hour the knot on the end of his dragging rope will collect a lump of mud as big as a washtub.

In some seasons great swarms of grasshoppers appear far overhead, glittering like hostile planes against the sun, swooping down to devour every green thing, leaving cornstalks and trees bare sticks to rattle in the burning wind. These hoppers, thick as Mormon crickets farther west, have only one thought—to devour everything in sight in the shortest possible time. They will even attack fence posts, barbed wire, upholstery, and—when all else fails—will devour each other. Fortunately they do not attack other living animals. Fowls and porkers thrive on them, and the farmer, left without a corn crop, thrives on the fowls and porkers.

In winter there are deep snows—and blizzards, which come on the wings of howling winds, filled with suffocating fine ice particles that freeze the lungs and blind the eyes. A man may get lost in that smother between his house and the barn and so freeze to death within a few yards of his own threshold.

In summer fearful thunderstorms sweep over the land—

LEWIS AND CLARK LEFT THE MOUTH OF THE MISSOURI I

MAY, 1804, RETURNED TO ST. LOUIS IN SEPTEMBER, 1806.

great black threatening clouds, the color of cut lead, accompanied by a furious wind and a deluge of slashing rain. Strangers are stunned by the bellowing artillery of the skies, amazed at rain which falls so hard that it bounces from the earth again, spraying a man to the knees. Such thunderstorms contribute a great part of the rainfall of the region. But of course water coming down at such a rate does not sink into the uneven soil; it runs off in torrents, washing away fertile acres.

These hazards are spectacular and dangerous enough. But to a man who has been through a hailstorm, they seem comparatively harmless and quiet affairs.

In the calm preceding such a storm, one has little warning. First a few scattered hailstones thud upon the ground. Clouds quickly darken the sky. Within a minute or two a volley strikes, turning the surface of the ground white as snow, pelting every creature not under cover.

Caught in such a storm, I once took refuge among some haystacks stowed under a sheet-iron roof. I heard a deafening rattle on the iron, which increased in volume and rose in pitch. My ears ached and my nerves were shaken by this shrill noise. Then the sound deepened to a hoarse roar until the stout posts supporting the structure shook under the furious blast of icy slugs. Immediately after a lake fell from the sky. Lightning flashed, but it was impossible to hear the thunder. Finally the scream of the storm sank to a rattle, the rain diminished, and I could see about me.

Everywhere on the hillsides, on the level, the country was covered with ice. Hailstones as big as hens' eggs—and bigger—tumbled and rolled in the flood of water sweeping them down the gullies and off the slopes. When the storm had passed, we found ice stacked up where the water had left it, and in one gully nearby to a depth of fully six feet.

Such a storm tears every shred of foliage from growing things, if indeed it does not beat them down to the ground.

Automobiles are dented as though they had been beaten with a hammer, and the tops are often broken in; roofs are damaged, windows broken, animals destroyed, and everything not made of iron or stone is shattered.

It is an ill wind that blows nobody good, and all such natural disasters make the Indians chuckle. If you know the old men well enough, they will nod wisely and grin and say, "The white man has taken our country, but he is paying for it. He thinks he is pretty smart, but he is not as smart as he thinks he is. The white man came into this country, cut down the trees, and turned the ground wrong-side up. He killed all the animals that would take care of themselves, so that he could bring in animals that he has to take care of. He calls this *civilization!*

"When I was a boy the water in the rivers was good to drink, there was plenty of timber in the valley, the grass was knee-deep everywhere, and there was always an antelope or a buffalo to be had for the killing. When I went on a trip, I did not have to take food with me; the Great Spirit provided my food.

"But now there is nothing to eat on the prairie. The grass is scanty; the trees are gone. When I drink the water of the rivers, it makes me sick—it even makes horses sick. And the wind has blown away the very earth from under my feet. All we have left is the weather."

That is the Indian view. But it has taken the white farmers two generations to begin to see the wisdom of it. Stubbornly they stuck to their homemade hell.

Most of the white people along the Upper Missouri have come into the country since the building of the railroads, and population there did not reach its level until after the turn of the century. Many of these settlers were recent immigrants who came to America full of hope and great expectations, to seize the opportunity of owning their own land, as they could not easily do in Europe, and to share in a standard of living higher than they had known in the old country. They had usually been pumped full of golden dreams by steamship and railroad companies, and sometimes arrived unprepared for the stern conditions of the frontier. Though they found something better than that they had left behind, things were never so good as they had expected; thus they had a considerable psychological adjustment to make, in addition to their other problems of adapting themselves to a new country, a new language, a new system of government, and new ways of making a living.

If it was tough for the ordinary American farmer who moved west to make a living plowing the Plains, it was far more of an ordeal for the small farmer, fresh from Europe, trained in methods that often proved inadequate.

These men lacked the restless, migratory habit of the American settler. For the American farmer has been, for centuries, nomadic as a miner. He regarded land as something to exploit rather than as a permanent investment. While things went well, he stuck; when they went badly, he moved on. There was always free land farther on, and owning property was no novelty to him.

But the European immigrant farmer, his heart warmed by the possession of a piece of land of his own, and with a family tradition of living in one community for generations, was much less willing to pull up stakes when the going got tough. He stayed and took it—as it seemed to the native American, unnecessarily. His stubborn efforts were heroic and his hardships very great. These have been reflected in a whole literature of the Northern Plains—a literature of earth-bound lovers of the land, grimly fighting it out with a harsh climate and a stubborn soil. Farming, to these people, was not an adventure or a speculation but a fight for survival, which has enabled them to contribute a toughness and strength to American life and letters that might have been lacking otherwise.

In Canada, on the plains north of the Upper Missouri, the people often seem discouraged by the disappointments and hard conditions there—perhaps because they are not aware that their experience was and is shared in great measure by all plainsmen clear down to Southwest Texas. But in Montana and the Dakotas, in Iowa, Nebraska, and Kansas, there is no lack of hope and laughter; the worst disasters are made into jokes. For though the ever-present threat of drought hangs constantly over their heads like the fear of an air blitz over European cities in wartime, the plainsmen go right on hoping for a crop, for better times, for a good year. They have faith in the country, some of them, even when they have not had a decent crop in nine years' time. As their saying goes: "Well, I have had three good crops: twelve years ago, five years ago, and *next* year!"

But this country east of the Missouri is *not* the "terrible country" which certain novelists have dubbed it. Novelists complain of "drab farms and monotonous towns," of "wretched architecture" and "forlorn vistas that nobody loves or could love." But it is *not* a land of despair, because the farmer has learned through hard experience that he must, like the Indian, observe the moods of Nature and prepare a simple remedy or means of escape. He has at long last learned that he must live *with* the country instead of fighting it, that he *must* work as nature worked to make that land so fertile. It is fortunate that the settlers there are such a hardy crew. They do not give up easily.

Last year North Dakota was shorthanded and nearly ten years behind in replacement of worn-out farm machinery—and the war made replacements impossible.

But, with good crops in sight, the North Dakota Plan, started in 1942, was again put in force. Doctors and lawyers, students and housewives, the butcher, the baker, and the candlestick maker all pitched in in their spare time, working in the fields to make good the lack of machines.

Meanwhile Governor John Moses got on the telephone and kept the wires hot until Washington sent out more than eight thousand soldiers and laborers to help with the harvest. In spite of all handicaps the state produced 154 million bushels of wheat—fully 10 million more bushels than Kansas, the nation's biggest wheat producer. Also, North Dakota produced 25 million bushels of corn, 65 million bushels of barley, 72 million bushels of oats, 22 million bushels of potatoes, to say nothing of a huge crop of flax. The state produced also half a million head of cattle, over a million head of hogs, nearly a million head of sheep, more than eight million pounds of wool, more than ten million chickens, more than fifty million pounds of butterfat, fifty million dozen eggs, and more than a million pounds of honey.

Threshing began, and the golden grain poured out in such a flood as soon filled every one of the tall, hunchbacked elevators studding the prairie. The state's own storage facilities—her mighty breadbaskets—were soon packed full of wheat, pressed down, shaken together, running over. Within a few days after the harvest started, the governor learned to his dismay that some eighteen million bushels of grain were piled out on the prairie where the first snow or rain might destroy it.

While the whole state held its breath, Governor Moses got on the phone again, sent frantic telegrams. The railroads hurried long trains of empties in to pick up the crop. By Christmas less than a million bushels remained exposed to the weather. That record crop had been saved.

For when the climate co-operates, the Upper River produces magnificently. In 1943 North Dakota harvested more food than ever before in her history. It is no wonder her farmers stick.

Nowadays they are learning to preserve the topsoil, letting all rubbish rot on the surface, as nature does, instead of plowing it under. They use a disk or bull-tongue plow to cut up the straw and vegetable matter instead of turning it under, so providing a protective mulch, making the most of the capillary attraction of the soil, and preventing the wind from blowing it away.

The moldboard plow which broke the Plains is being thrown on the scrap heap. The farmers are turning their old plowshares into swords, and with the new ones cultivating the soil instead of gutting it. They are building windbreaks and laying out their fields so as to hold all the rain that falls. Every smart farmer up there carries in his pocket a copy of his gospel, Ed Faulkner's *Plowman's Folly*. He knows now that the irregular order of nature, though not so pretty perhaps as the long, straight furrows of our fathers, is the only *real* order—the *only* order which will produce without destroying the soil.

On bottom lands along the river one sees rich, bright carpets of dark-green alfalfa and gray-green fields of tall hay; timber and dense underbrush bind the banks of the stream. The hills are studded with cattle, and marked by lines of taut fencing.

Towns are few along the stream on the east bank. Indeed, ruins of forts and agencies and Indian villages are more numerous than inhabited places. But on the west bank there is even less to show for man's occupation. Groves and roads and even ranches are rarely within sight from the river. There the farms end and the cattle begin.

It was not always so. Once, not so long ago, there were cabins and shacks and sod houses, fences and trails stretching from ridge to ridge, cutting square patterns into the scanty grass. Within thirty years all this has given way to ranching. The covered wagons with their ribbed tilts pulled out eastward across the river through clouds of dust; the defeated movers headed "back home." There is still some farming where the soil is suitable, but for the most part only forage is grown. Not so far away beyond the rolling prairies are the buttes and badlands where nobody could hope to make a living with a plow. It is a long, long way from the Missouri River west to the rich irrigated region about the Black Hills.

No man raised east of the Missouri can understand the feeling of the plainsman toward rain. It is the answer to his prayer and puts a lump in his throat or grinds out the arid humor from his heart. But on the Northern Plains, there is the added threat of winter. Even in midsummer, the whole country seems to hug the ground. Weather means success or failure. As men say there, "We look into the sky to see where our winter underwear is coming from."

The first white men to inhabit the Upper Missouri were fur traders, hunters, and trappers, many of whom went there to escape from civilization. Among them was a certain number of sentimentalists and idealists, including artists and writers who poetized and sentimentalized the West. But even the hard-boiled go-getters who came there to make money, who despised the Indians and grudged the discomforts of the frontier, also bitterly hated the cities and towns from which they had come. They wanted no settlers coming in to kill off the game and destroy their monopoly.

The ranchers who followed the fur men were quite as hostile to eastern ways. They, too, did what they could to keep settlers out—partly from motives of self-interest, but fundamentally because of their love of the life they led.

In the old buffalo days and the later days of the open cattle range, main highways on the Plains ran north and south between the mountains and the Missouri. It was only when the plowmen and the gold seekers trekked westward that the old north-and-south trails were neglected and the main highways ran east and west. This reversal of the natural movement on the Plains was at once a cause and a symbol of the new culture sweeping in. The wagon trains knocked the old culture dizzy. Finally the railroads slew the traffic on the Missouri and on the north-south cattle trails.

South of the Santa Fe Trail in Texas and Oklahoma, this was much less true. The early development of natural resources there and the healthy agriculture in that southern country created a more self-sufficient community, which has never been entirely dominated by outside influences. There people still think in terms of north and south rather than of east and west. A town down there thinks of itself as the end of the trail—not as a way station on a transcontinental highway.

By contrast, the Northern Plains sometimes seem overwhelmed by outside influence. Chicago and St. Paul dominate the Dakotas and Montana to a far greater extent than any towns outside Texas or Oklahoma can do there. In the north one sometimes finds a sort of resigned provincialism, a kind of acceptance of the authority of outsiders to the east.

Perhaps the Canadians in that country have had something to do with this state of mind. For in Canada the Mounties and the railroads moved in before the settlers were numerous, and moreover your Canadian is always proudly conscious of British standards in his background. Even the fur trade in Canada was owned in, and operated from, England. The Hudson's Bay Company was, in fact, a government in itself. However it came about, the plainsman in the Dakotas and Montana, though self-respecting and standing on his own feet, seldom exhibits the exultant arrogance commonly encountered in West Texas. He does not go in for big hats and boots unless they are necessary to his business.

It is true that the dude ranch in the north has made the Old West pay, but there dude ranches, unlike those in Texas and Oklahoma, are patronized by Easterners and even financed by eastern money. To the native, it is not a spontaneous expression, but a kind of antiquarian revival. In Oklahoma, on the contrary, even the Indians have not been entirely overwhelmed by outside influences. Though they have kept up the Sun Dance, it is not as a spectacle for white men. Though Sun Dances are given in Oklahoma every summer, not one white man in a thousand in that state has witnessed one. Very few know that these ceremonies are still performed.

The Sioux Sun Dance was suppressed in 1882. But recently a Chamber of Commerce in South Dakota realized that there was tourist money in it and induced Chief One Bull to revive the ceremony.

Some plainsmen in the north wish that the Missouri had been the dividing line between the Dakotas, so that they might have an East and West Dakota instead of a North and South Dakota. They think that the plainsmen might hold their own better if they were not overbalanced by a larger and wealthier population on the farms east of the river. However, it might be a bad thing for national unity if every state were a cultural and economic unit.

Nevertheless, there is an admirable side to this dependence upon remote centers, namely, that the plainsmen of the Upper Missouri country seem much readier than the Texans to welcome the cultural contributions of aliens. The rapid development of natural resources on the Upper River has attracted great numbers of immigrants from Europe—many from the more advanced nations—who have brought with them valuable traits and ideas and customs. These have been made welcome.

On the Upper Missouri there are still French and French Indians and Scottish people descended from the fur traders; great numbers of Canadians, a large contingent of Germans, and even more Scandinavians, besides many from the British Isles and the Irish Free State, and Russians, Finns, Yugoslavs, Italians, and Czechs, with a much smaller number of

Mexicans, Negroes, and Orientals. These comparatively recent arrivals appear to play a larger part in the life of the Northern Plains than do similar new elements on the Southern Plains. And this cannot but enrich life there. Though these people also may still feel strongly that they are derivative, their cultural and economic and sometimes even political roots lie far away. This in some measure frees them from immediate domination by the cities in adjoining states.

The Missouri River welcomes hardy strangers. That stream offers an inspiring challenge to men of vigor, calling them from afar. In fact, a large proportion of the men whose names are famous for their connection with the Big Muddy came from other regions, and, often enough, left the valley when their vigor waned. That river has always been the home of men with hair on their chests.

Of these, few were more hirsutely chesty than the Mountain Men. . . .

Outpost

CHAPTER XVI

Mountain Men

IN 1822 the fur trade took a new turn. Up to that time traders had operated from fixed forts, buying peltries from the Indians after the manner of the Hudson's Bay Company in the British possessions. In Canada the Indians farmed out the wilderness—each band or family occupying or hunting on a definite range, and the factors at the trading posts managed things so that the Indians did not exterminate the game. Under their monopoly the fur trade might have lasted forever with that system.

But in America the spirit of independence and every-man-for-himself, and the absence of any adequate authority made such a system impossible. On the Missouri River, moreover, the fur traders found themselves dealing with Plains Indians, nomads and marauders, who had no notion of sticking to a definite hunting range if they could find better hunting on the lands of their neighbors. White traders, wandering over the country on their own, also had no regard for tribal boundaries. It was apparent that the British system of managing the fur trade would never work south of the international boundary.

The Plains Indian by temperament and necessity was neither a hireling nor a laborer. He was a hunter and a warrior. He did not like payments, he preferred gifts.

The Spanish and French officials had understood this, and their governments had seldom attempted to buy land from the Indians or pay them for their services. Instead they followed Indian custom, making feasts and presents when they wanted a concession and helping the Indians out with gifts when they were hungry or in trouble. That was why they had so little trouble with the very tribes who later terrorized the country after the Anglo-Americans moved in.

Major William H. Ashley and his partner Major Andrew Henry, considering these facts, devised a new plan for harvesting the fur. Instead of building and maintaining permanent forts at great expense to trade with Indians, they proposed to hold an annual fur fair or rendezvous in the mountains at whatever point was most convenient for their white trappers. They proposed to carry out to the rendezvous goods to be traded for the furs their trappers might bring in. In this way their headquarters would always be where the fur was thickest and their trappers would never need to return to the settlements or make long marches to a distant fort, but could keep busy the year round and live always in the Indian country. This meant more fur, more income for the trapper, and far less expense for the bourgeois.

This system kept the trappers in the mountains for years at a time and produced that doughty breed to which Kit Carson, Jim Bridger, Tom Fitzpatrick, Hugh Glass, Jedediah Smith, and John Colter belonged—the Mountain Men.

It was launched in 1822. On March 20th the *Missouri Republican*, a newspaper published in St. Louis, carried Ashley's advertisement addressed "TO ENTERPRISING YOUNG MEN," announcing that the major wished to employ "100 young men to ascend the Missouri River to its source, there to be employed for one, two, or three years."

Within a month Henry and Ashley had got a license to enter the Indian country. They embarked their men in two keelboats of unusual size—each more than a hundred feet long. The cabin, or cargo box, was eighty feet long and there was a galley forward of the mast for the cook. The handle of the great sweep at the stern rose to the top of the cabin, where the steersman, or pilot, had a clear view up and down the river. Like other keelboats, these had oars, poles, and a square sail on a mast forward of the cargo box.

They "set poles for the mouth of Yellowstone," and though one of the boats sank near Fort Osage on the Lower River—a loss of some $10,000—the other boat fought its way up to the Yellowstone, where the partners built their fort.

The next year, 1823, the partners hired another hundred men and launched another pair of keelboats. They called these the *Yellowstone Packet* and the *Rocky Mountains*. Ashley wanted to trade with the Arikara, or Ree, Indians, whose villages were on the river near the mouth of Grand River in South Dakota. The two boats tied up below the villages there the last day of May.

In those days, before the great epidemics of smallpox and cholera had decimated the tribes on the Upper River, Indians there were thick as fleas, and the Rees were among the boldest and most warlike—as they had to be to stand up to the far more numerous Sioux. In 1805, when visited by Lewis and Clark, the Rees had been friendly enough. They were fine big men and stubbornly clung to their own customs, telling white men who offered them whisky that they were amazed that their "fathers" would give them anything like that to make them foolish. But by 1823 they had seen enough of white men and had enough trouble with them to be ready for mischief.

Unscrupulous fur traders had encouraged the Indians to murder their rivals in the trade, with the result that the Indians had come to despise all the whites and regard them as possible prey. There had been killings and horse stealings ever since the days of Manuel Lisa. Ashley's men, though warned by Joshua Pilcher at the trading post of the Missouri Fur Company below, took few precautions. They accepted the Rees' professions of friendship at their face value. Pilcher himself had had trouble with the Rees because his men would not give up to them some Sioux who were traveling with them. He knew that the Rees would never rest until they had got revenge.

The Rees wanted Ashley to trade inside their palisaded village on the curving bluff, but the major replied that he would trade on the bank of the river opposite his boats, which were anchored in the middle of the stream a hundred feet or so offshore. Ashley traded powder and ball for horses, since he proposed to send forty of his men overland to the Yellowstone.

The trading went on without a hitch. But Edward Rose, the half-breed whom Ashley had employed as interpreter, and who had lived in the Ree village for years, had a hunch that trouble was in the wind. The chiefs were too polite, the warriors too silent, the women and children too shy. He kept after Ashley to get out of there while the getting was good.

Rose, however, was not a man whom Ashley could like. He inspired distrust in Ashley as he had in Hunt in 1811. Rose was, as Irving writes, "a dogged, sullen, silent fellow with a sinister aspect and more of the savage than the civilized man in his appearance." Just what Rose could expect to gain by misleading Ashley is not clear. Nevertheless, Ashley would not listen. The fellow's record as a river pirate and member of a chain gang was against him.

The Rees invited Ashley to a feast, and Ashley went to dinner with the chief in his great domed earth lodge in the nearer village.

It was a big town with some seventy earth lodges, like miniature mountains dropped helter-skelter on the bluff.

These stood within a palisade a dozen feet high made of vertical logs a foot in diameter, with breastworks and a dry ditch. So Ashley went along and ate and smoked with his entertainers. They let him go away unhurt. They wanted bigger game—the whole party.

On his return Rose warned Ashley again; but it was no good.

The Indians had fortified their position cleverly, and Ashley had anchored his boats within the trap. Instead of passing the villages before he tied up to trade, he had anchored opposite the lower village in a narrow channel between the bluff and a great sandbar on which the Indians had built a fort of logs to command the channel.

Even when one of the chiefs from the village slipped through the darkness to warn Ashley, the major took no action. He thought the best way to prove his friendship was to appear careless of danger. Such conduct only made the Rees think him a coward and a fool—and, what was worse, an easy mark.

Then, to make temptation irresistible, Ashley left the horses he had just acquired on the beach below the village with a guard of only forty men.

Ashley climbed into his bunk on the keelboat and slept soundly until nearly four o'clock. When he woke up, he found Edward Rose shaking him. Rose reported that the Rees had already slipped arrows between the ribs of one of his men and killed him; the Indians would attack at dawn. Now, at last, Ashley realized the danger in which he stood and began to think of defense. But before he could do much, the Indians opened fire from their palisades on the bluff.

The two villages counted some seven hundred warriors. Most of them had guns, as Ashley well knew, having supplied them with powder and ball only the day before.

Those guns were popping all along the line—a line at least six hundred yards long.

The Indian fire was concentrated on the horse guard. The men, lying on the open beach, suffered severely. Those who survived the first volley "forted" behind the carcasses of the horses which had been killed. Ashley, hoping to save them, and the remaining horses too, yelled across ordering the men to swim over with the animals to a shoal in the middle of the stream.

They tried to do it. But so many were hit that Ashley ordered them to go back.

He then appealed to his boatmen, telling them to man the skiffs and row to the beleaguered men on the beach.

The boatmen, not having been hired to fight Indians —much less to paddle around unarmed under heavy fire— had no idea of obeying such an order. Suicide was not in their contract. To a man they mutinied. Ashley, instead of pulling a gun on them, let the matter drop.

But the enterprising young men on the beach were full of fight. They proceeded to load and fire with a steadiness that would have done credit to veteran soldiers. They had the stuff that later made many of them famous.

Ashley's whole thought was to get his men aboard. Pleading and bribing, he at last persuaded a few men from the keelboats to row the skiffs ashore. But now the men ashore, having lost comrades in their first attempt to cross the water, could not be induced to leave the beach again. Only the wounded came back to the keelboats.

Still Ashley persisted. He sent one skiff over for the second time. This time the Indians let the two boatmen have it. One was killed; the other took cover in the bottom of the boat. Unmanned, the skiff floated away downriver. After that, none of the men would venture from the keelboats.

By that time all the horses had been shot down and

half the men on the beach were wounded or dead. The Rees crawled forward to the edge of the curving beach, firing into the whites from three sides. The position was too hot to hold. At a word, the survivors rushed for the river, splashing and swimming for the boats, while the bullets kicked up the water about their heads. A few died on the beach, some were swept under by the water and drowned, or perhaps were killed by Indians waiting below. The icy water accounted for some of the wounded. But others managed to make it to the keelboats and crawled on board. The boatmen cut cables; the boats drifted down out of range.

Reed Gibson was mortally wounded. Thirteen others had been killed. There were eleven wounded, including old Hugh Glass. The disaster appalled Ashley. Afterward he claimed that, if his orders had been obeyed, he "would not have lost five men."

But Ashley was in command; it was his business to get his orders obeyed. If he had gone upstream past the villages as soon as his trading was over, or even dropped down below the villages to make camp, he might have saved his men. Instead, he had attended a feast in the village.

Next morning he proposed that they all go upstream again.

His men thought he had lost his head completely. There was no wind to carry them past the Indian towns, the current was too swift for oars, perhaps too deep for poles. Nobody cared to tow the boats up that channel under the guns of the Rees. The boatmen openly refused to go; a good many of them made up their minds to desert. Ashley, surprised and mortified, and afraid that if they went they might take his boats with them, called for volunteers to stay with him until he could get in touch with his partner, Major Henry. He now had some eighty men left. Of these, thirty, including five boatmen, volunteered to stay.

Ashley transferred the cargo of the *Yellowstone Packet* to the *Rocky Mountains,* put his wounded men aboard and sent the boat to Fort Atkinson—450 miles below—with a letter to Colonel Henry Leavenworth urging him to bring his regulars to whip the Indians. It is probable that Ashley did not really expect Leavenworth to act, since the colonel got his orders from St. Louis. Ashley's real hope lay in relief from his own fort at the mouth of Yellowstone. He determined to send word to Major Henry.

Ashley would have gone himself, but he dared not leave his boats. He had to hold his men together, save his cargo, and so keep himself from ruin. His handful of discouraged men could not safely be divided. He could spare only one or two. He asked for a volunteer.

The boatmen knew that Ashley did not expect any of them to go. The others listened to his pleas in sullen silence. They felt that they had already risked their skins once too often for their stupid employer. Others may have hoped that no one would go, which would relieve them of their promise to wait for Major Henry. It was all of three hundred miles as the crow flies to the mouth of Yellowstone—450 by the river—and every mile a gantlet through a country swarming with hostile redskins. Nobody budged.

Those silent minutes were bitter for Ashley. He looked at his men, and they looked back at him. He knew what they thought of him now. Ruin stared into his red face.

All at once the youngest man in the party spoke up— Jed Smith. "I'll go." Jedediah Smith was a lank young fellow with steady eyes, moccasins spattered with blood from yesterday's battle, and a Bible sticking from the pocket of his hunting coat. His buckskin breeches bagged at the knees, because Jedediah knelt down every night to say his prayers. He had shown courage aplenty on the beach and he had brains. A York State man, well educated, of good family, it was his first time in the mountains.

MOUNTAIN MEN

The men admired the young 'un's spunk, though none of them offered to go with him. But Ashley went to work on a French Creole who knew that country, and finally talked him into going with the boy. After dark they set out.

When Major Henry got that message, he lost no time. He boarded his boats with all but twenty of the men in the fort and shot down the Missouri as fast as he could go. He overtook Ashley at "the Fork"—the mouth of Cheyenne River—on July 1st. Soon after, the partners rejoiced to learn that Colonel Leavenworth with six companies of the Sixth United States Infantry was close at hand and marching up the river to their aid. Three keelboats carried his supplies and artillery.

Ashley's defeat at the Ree villages had given Joshua Pilcher, president of the Missouri Fur Company, plenty of worry. So, now, he too was delighted to learn that Leavenworth was coming. The Rees had killed some of Pilcher's own men, and he knew that unless the Rees were soundly whipped, there would be no end to trouble along the river. Pilcher asked the Indian agent, Major O'Fallon, to appoint him subagent, and with this authority got five hundred Sioux warriors to join forces with the troops. The Sioux were delighted to help wipe out their ancient enemies, the Rees. They knew the Rees for tough, mean warriors—men who would even plunge into the river and kill swimming buffalo with knives!

But just then the Missouri River saw fit to take a hand in affairs, and with a snag sank one of the colonel's keelboats. He lost seventy muskets and seven men, to say nothing of the rest of the cargo of supplies.

This disaster appears to have cooled Colonel Leavenworth's ardor. He had not waited for orders from headquarters, but had set out on his own, expecting to cover himself with glory. Now he had lost one-third of his supplies—supplies for which he would be held accountable. The more he thought about it, the more he repented his rash action. His enthusiasm for the expedition steadily waned.

Still, the colonel did not feel like turning back without striking a blow.

He went on to Fort Recovery, where the expedition was reorganized as the Missouri Legion.

The fur traders were glad to provide rifles in place of the muskets lost in the river. The companies of regulars, the trappers under Ashley and Pilcher, and the five hundred Sioux under Chief Fireheart combined forces and marched upstream. Leavenworth gave temporary army commissions to some of the mountainmen. Jedediah Smith, having demonstrated his judgment and courage, was made captain; William Sublette, sergeant major; Tom Fitzpatrick, quartermaster; and Edward Rose, now restored to favor, was made ensign. Angus MacDonald and Henry Vanderburgh were commissioned captains. William Gordon was made a shavetail and Moses B. Carson a first lieutenant. The expedition now included almost eight hundred men.

With the Sioux nothing succeeds like success. When the Legion reached the Ree villages on August 9th, three hundred more Sioux came riding to join the party, burning with martial spirit. Leavenworth marched against the towns.

Pilcher's outfit went above the villages, to head the Rees off in case they tried to get away; while the Sioux, unable to wait for the troops marching up the riverbank, piled on their war ponies and dashed to the attack.

The victorious Rees, full of fight, rushed out of their towns and pitched into the Sioux. Dust and powder smoke filled the air, as the Indians, yelling and shooting, charged to and fro after their savage fashion.

The Rees were in no mood to retreat. Even when the Sioux killed thirteen of them, the rest stood fast. It looked to Pilcher as if the Sioux might get the worst of it. He hurried back and called for reinforcements.

On came the Legion: Ashley on the right next the river, the regulars in the center, and Riley's riflemen on the left. When they reached the Sioux line and opened fire, the Rees lost heart and hurried back to the shelter of their fortified villages. Colonel Leavenworth decided to wait until the boats brought up the cannon. The battle was over for the day.

If anyone had expected the Rees to pull out during the night, he was disappointed. Colonel Leavenworth split his command in order to attack both villages simultaneously. His cannon balls crashed into the town, killing Chief Gray Eyes and knocking down a medicine pole. But the Rees soon discovered that the small cannon balls could not destroy their palisades. They merely kept their heads down and went on shooting. The colonel, not wishing to make a frontal assault on the Ree entrenchments, deployed his men and kept them firing at random into the villages.

This sort of thing bored the Sioux. Long-range shooting was no part of their tactics. They begged Pilcher to rush the towns, force the Rees into the open, and so settle the business quickly. But Colonel Leavenworth shook his head. He went on banging away until all his cannon balls were gone. Then he led his men back downstream, where they made a meal of the crops in the Ree cornfields. It was the first meal they had had in forty-eight hours.

By this time the Sioux were contemptuous. They had done all the fighting that amounted to anything, had lost several men. The white men had suffered no loss and accomplished nothing. Fireheart declared that the Americans were just "a heap of squaws."

The mountain men were outraged. They told the Indians that in the fight next day they would show them what warriors Americans were.

No sooner had they made their brag, however, than Colonel Leavenworth proposed to make peace with the Rees!

Pilcher was fit to be tied when he heard that. While

the colonel and the Ree chiefs were passing the pipe, he paced back and forth, back and forth, like a warrior itching for a fight. Seeing this, the Rees took alarm. They told the colonel that they would *not* make peace with him unless Pilcher and his mountain men were a party to it. Then the colonel's face went red.

Fireheart and his disgusted Sioux now washed their hands of the whole business. They pulled out and lined up on a ridge nearby. The chief sent word to the colonel that he and his warriors would look on and see which side won out!

The contempt expressed by everybody for the colonel only made him the more stubborn. He pulled his rank on the mountain men, reminding them that they were under *his* orders. "You will do as *I* say," he warned them. Nothing the mountainmen could say had any effect on the colonel. They realized that, unless the Indians were soundly whipped, the lives of all white men on the river would be in danger for years to come. But the colonel insisted that *he* was in command, and that the interests of the United States required a truce.

The Rees, having no confidence in the colonel, distrusting the mountain men, and probably fearing that the Sioux would join in after all once the fight began, asked the colonel to postpone his attack until the following day. And Leavenworth agreed!

Of course, when the sun came up the next day, the villages were empty. Leavenworth marched up the hill and found one lone Indian—an old, blind woman who had been left behind. He gave her rations and then marched down again. She was the only one to receive any benefit from Leavenworth's costly maneuvers. The command headed downriver.

Some of the angry mountain men, determined to injure the Rees, sneaked back and set fire to the Indian towns.

Leavenworth had the bad judgment to reprimand Pilcher for that. Pilcher turned on him and gave him such a tongue-lashing as no other officer of the United States Army ever had to put up with. And when the colonel explained that he had not meant to blame Moses Carson and Henry Vanderburgh for the fire, the two of them, their angry eyes full of tears, protested to their comrades that they had never done *anything* to deserve the praise of such a man. Pilcher, not content with his tongue lashing, wrote a letter to the colonel:

"Humanity and philanthropy are mighty shields for you against those who are entirely ignorant of the disposition and character of Indians, but with those who have experienced the fatal and ruinous consequences of their treachery and barbarity these considerations will avail nothing. You came to restore peace and tranquillity to the country, and to leave an impression which would insure its continuance. Your operations have been such as to produce the contrary effect, and to impress the different Indian tribes with the greatest possible contempt for the American character. You came (to use your own words) to 'open and make good this great road'; instead of which you have, by the imbecility of your conduct and operations, created and left impassable barriers." [1]

So these men hated, despised, and reviled each other, and nobody thought to lay the blame on the Missouri River, which had sunk the colonel's keelboat and so caused all the trouble.

The river, of course, had nothing to say. It went right on with its work, planting snags to change the course of history.

CHAPTER XVII

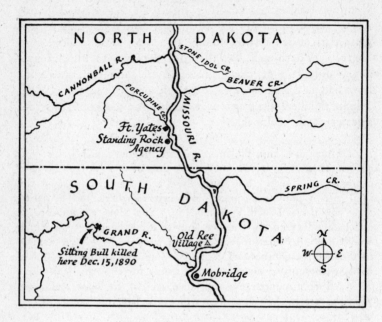

Standing Rock

As one mounts the Missouri from South Dakota toward the mountains, the course of the river cuts through country ever more rocky, and many of the landmarks within sight of the water are of stone. The best known of these in North Dakota is the Standing Rock, which has given its name to the Standing Rock (Sioux) Agency.

The agency and the military post, Fort Yates, stood close together on the bluff overlooking the river's west bank.

STANDING ROCK

About a quarter of a mile northwest of the agency lies an open slope, known as Proposal Hill because romancing officers and their ladies used to go there to spoon. Hereabouts, when the agency was first established, the white men found *Inyan Woslata*, the Standing Rock.[1]

Indian legend tells how, when a certain brave took a second wife, his first wife became jealous. When the camp moved, the young woman refused to stir, and remained sitting by her lodge fire with her child on her back and her little dog by her side.

The tribe moved on, expecting her to follow. But when she failed to show up, her husband became alarmed and sent his brothers-in-law to find her. Going back to the abandoned campsite, they found their sister, with her child and dog, still seated by the fire, all turned to stone. After that, the Indians carried the rock with them on their wanderings as a sacred object. Certainly the stone does not belong on Proposal Hill. It is of a dark-gray metamorphic composition quite alien to the region.

All that country once belonged to the Rees, and the Sioux say that they acquired the stone and the legend along with that country from the Rees.

It is possible that it once stood on that small stream across the Missouri just above Grand River which Lewis and Clark dubbed Stone-idol Creek. Sometime during the eighties James McLaughlin, the agent, moved the stone to the agency and set it up on a pedestal overlooking the river, where it may still be seen. Unhappily the little stone dog was built into the pedestal by mistake and is no longer visible.

The Standing Rock or, more correctly, the Upright Rock, somewhat resembles the figure of a diminutive seated woman wearing a robe or shawl.

The Standing Rock Reservation lies on both sides of the North Dakota-South Dakota boundary, and after the

SITTING BULL

surrender of the Sioux was the home of Sitting Bull, Gall, Running Antelope, John Grass, and Rain-in-the-Face. A good many stirring events took place on the reservation, and there Sitting Bull was killed and buried.

It was at the Standing Rock Agency, also, that Tom Custer had his struggle with Rain-in-the-Face.

During the Yellowstone campaign several civilians were attached to the military expedition. Two of these, the sutler and the veterinary surgeon, had a dangerous habit of riding off unarmed together and straggling behind the column.

One day when these two men stopped to water their horses, they were fired on by some Indians from a coulee within sight of the regiment. When the bodies were recovered, there was a good deal of indignation at the "murderers."

Of course, to the Indians, any white man able to bear arms and traveling with a military expedition was, for all practical purposes, an enemy. It is not likely that Rain-in-the-Face knew anything or cared anything about the white man's distinction between civilians and soldiers. But the white men did not forget.

More than a year passed, and the Seventh Cavalry was back at Fort Abraham Lincoln, when news came that Rain-in-the-Face had visited the agency at Standing Rock to draw his rations and ammunition and had publicly boasted of killing the two men. He had even acted out his bloody deed, to show how he had beaten one of them to death with his war club.

Immediately a detachment of a hundred men and two officers was ordered up the river. General George Armstrong Custer knew very well that, if the Indian scouts at the post learned of the object of the expedition, his purpose would leak out and reach Standing Rock ahead of the troops.

Accordingly he gave his brother, Colonel Tom Custer, sealed orders, so that no one would know that his mission was to capture Rain-in-the-Face. The force arrived at Standing Rock on ration day, when hundreds of Indians were at the agency—all of them, as usual, carrying rifles. When the Sioux saw the troopers, they immediately became suspicious. But Tom Custer fooled them by sending a part of his force to a camp ten miles distant to make loud inquiries for three Indians who had committed some crime on Red River. Meanwhile the colonel took five picked men and strolled into the trader's store on the edge of the bluff. There he and his men lounged about waiting for their man to show himself.

It was a very cold day, and the Indians were wrapped in their blankets. The Plains Indians had a thousand ways of wearing a blanket, and not infrequently contrived to cover everything but their eyes, like a Moslem woman. On such an occasion, when they were suspicious because of the presence of the troops, few of them cared to expose their faces. Colonel Tom couldn't tell one Sioux from another.

Time dragged on. But finally one of the Indians loosened his blanket—and the colonel had found his man. Drifting behind the tall Indian, he suddenly threw his arms around him. Rain-in-the-Face tried to cock his Winchester, but Tom was too quick for him. Probably Rain-in-the-Face expected to be killed at once, but he showed no fear. Indeed, it is quite possible that he went into the store on purpose to show his fearlessness and had loosened his blanket in bravado. Quickly the soldiers bound his hands and mounted guard over him.

The Indians swarmed around; an old man began to harangue the warriors. It was a tense moment.

But the captain with the troops was an old hand with Indians. Through an interpreter he explained that Rain-in-the-Face would receive a fair trial, that the troops would never give him up alive, and that the chiefs had better call off their young men.

Some of the relatives of Rain-in-the-Face then proposed that two other Indians be given up instead of the guilty man. His family was large and contained several influential leaders, notably Iron Horse.

In spite of all this, Tom Custer took his prisoner out of the store and marched him off to the military camp. The Indians all scattered and went back to camp—to get their horses. Not long after, fifty mounted warriors rode at a gallop through the agency to take position on the colonel's road back to Fort Abraham Lincoln.

In spite of these warlike demonstrations, no attack was made; and Tom Custer delivered the prisoner to his brother, the general, as ordered.

General Custer questioned Rain-in-the-Face at length, and finally persuaded the prisoner to confess his crime or—perhaps more accurately—to brag of it.

Rain-in-the-Face told how he himself had shot the older man, who soon fell from his horse, and how he had beaten his brains out with a stone war club and then shot his body full of arrows.

The other white man, he said, had signaled from the bushes where he had taken refuge and tried to make peace. Knowing that he was unarmed, the Indians closed in. The doomed man gave Rain-in-the-Face his hat as a peace offering, but the Sioux shot him with a gun and then filled his body with arrows. One of the arrows, entering his back, passed clean through and protruded in front. Rain-in-the-Face said that the dying man tried to pull it on through. He explained that he had not scalped the white men because the old man was bald, and the young man had had his hair cut short.

Naturally Iron Horse, the brother of the prisoner, had followed him down to the fort and brought him some fine clothes to die in. In the presence of the offcers he waited until Rain-in-the-Face was brought in, his chains clanking and with the guard at his heels. Rain-in-the-Face wore one black-tipped eagle feather erect in his hair, black woolen leggins and breechcloth, and covered his naked torso with a blanket of the same color, having a broad band of beads across it.

Iron Horse, supposing that Rain-in-the-Face would be hanged immediately, now exchanged his handsomely embroidered buffalo robe for the dingy blanket his brother wore, and also swapped pipes with him, so that Rain-in-the-Face might present the more handsome one to the general. On hearing that it was likely Rain-in-the-Face would be tried in Washington, D.C., Iron Horse took off his silver president's medal and hung it on the neck of his brother, probably in the hope that this would count for the culprit when brought before the Grandfather at the White House.

The guardhouse at Fort Abraham Lincoln was not a secure prison. A white man (who had been caught stealing grain) was imprisoned there with Rain-in-the-Face. The two were chained together.

This white man, being more ingenious than his fellow prisoner, managed to break the chain which bound them together, cut a hole in the wall, and, with the help of friends outside, escaped. Rain-in-the-Face made haste to follow him and scuttled off to join the wild Indians with Sitting Bull.

From there he sent a saucy message to Tom Custer, threatening to get even.

When Colonel Tom Custer was killed on the Little Big Horn battlefield, it was claimed that Rain-in-the-Face cut his heart out and ate it.

There is no truth in this legend. Rain-in-the-Face was very much overrated; his lameness, for example, was not caused by a wound received in battle, as the white men supposed. He was hurt in a buffalo hunt.

But in later years people regularly tried to pump Rain-in-the-Face about his alleged adventures. Among these interviewers, they say, was Frederic Remington, the painter.

Now Rain-in-the-Face, they say, loved the bottle, and finding Remington so eager to listen to his tales, made it clear that he would settle for nothing less than a quart of whisky. Of course, it was strictly contrary to regulations to give liquor to an Indian, but historical research has covered a multitude of sins. In his own writings [2] Remington declares that he had given liquor to an Indian when seeking information, so we may suppose that the tales told about him and Rain-in-the-Face have some basis in fact. Remington always hesitated to violate the law, they say, but Rain-in-the-Face knew how to force the white man's hand: he would walk out into the perfect privacy of that open country, sit himself down on Proposal Hill, and wait. Sooner or later the portly painter would come puffing up the hill, bottle in pocket, quite unable to resist the implied invitation.

Now the trouble with Rain-in-the-Face as an informant was that he got most of his facts out of the bottle. He would not talk until he had emptied it; and after he had emptied it, they say, he never told the truth. However, it is probable that Remington cared little how veracious Rain-in-the-Face might be. The bigger the lie the more interesting he found it. Certainly he gathered some fine Indian tales from *somebody*—and maybe it was Rain-in-the-Face.

From the time of Remington and Rain-in-the-Face, almost two generations have lived and died at Standing Rock. During all that time the Chinese cook who ran the hotel remained an institution, unchanged and unchanging as the rock itself. He was a good cook, but served one unvarying meal to all comers three times a day for nearly fifty years. No one with the price and the patience to eat there ever went hungry at Standing Rock.

But it was not always so before the days of that Chinese cook.

It was just here that the Cheyennes experienced a famine when there was scarcely a day's provision left in the camp. The situation was so urgent that the chiefs met in council to decide how their limited rations should be apportioned. When the council was over, the eldest man announced their decision.

"The young hunters are our only hope of survival; let them eat first and go looking for buffalo. The young women are the future mothers of the race. They must be saved; let them eat next. The children are the future men and women of the tribe. If they die, they can be replaced; but we do not wish to lose them. Let them be fed last.

"We older people have lived our lives. All we can contribute to the future of our nation is the food we can do without. We shall not eat at all."

Famine was an ever-present menace to Indians on the Upper Missouri until they obtained horses from white men. Some of these reached the Upper River early.

CHAPTER XVIII

Four Bears

IN THE YEAR 1738 Pierre Gaultier de Varennes, Sieur de la Verendrye, came down from Fort La Reine in Canada and reached the village of the "Mantannes" on the Upper Missouri. This was the first recorded visit of a European to the sedentary Village tribes on the river. After that there were many, and most travelers made some stay in the palisaded towns of the Mandans, Hidatsa, and Rees—partly because they were easily accessible and always at home, and partly because hospitable trading posts were commonly close by.

But there were other reasons. Lewis and Clark had spent the winter in their triangular Fort Mandan near these villages, and had time to give a considerable account of their manners and customs.

Scientists and artists followed one another up the river almost as regularly as migratory waterfowl, until the Village tribes were as well-known as any on the continent. A good many people in those days believed that the American Indians—or, at any rate, some of them—were descended from the Lost Tribes of Israel. The fact that the Mandan village contained quite a number of children and young people with light hair and eyes encouraged this supposition. The light complexion of these Indians might better have been explained by the fact that numbers of white men had resided in the neighborhood and even in the villages themselves, but pious people preferred not to think of such things.

The Mandans received more of this interesting publicity than did the other two tribes.

They, like the others, lived in villages of large round earth lodges, each of which housed several families and might be as much as sixty feet in diameter inside. To build such a lodge, the Indians first excavated a circular space the size of the proposed building about two feet deep, and made a slightly deeper hole in the middle to contain the fire. About this central fireplace, halfway to the edge of the excavation, four stout forked poles were set up, one at each of the four quarters. These pillars might be a foot in diameter and ten or fifteen feet high. Stout beams were then laid in the forks of the pillars, forming a square.

At the edge of the excavation, similar shorter posts five or six feet high were set up at regular intervals and connected in like manner by placing beams in the forks all around. Against this outer series of beams, poles were laid from the ground outside slanting inward and upward to form the walls. Then long rafters were laid from this outer wall toward the center of the building, slanting upward upon and beyond the beams supported by the four main pillars about the fire. The ends of these rafters were then trimmed off, leaving a smokehole in the middle of the roof about the size of a washtub.

Having completed the structure, the Indians now covered the roof and walls outside with brush and grass and topped these with sod and earth to a depth of several feet. The door of the building opened outward upon a narrow, flat-roofed vestibule perhaps six or eight feet long. The house looked like a gigantic mud igloo.

The floor of the vestibule sloped upward to the level of the ground outside. The door was closed by means of a hide curtain stretched on poles.

Around the walls inside, the beds of the inmates were arranged. Every bed was raised above the ground, and boxed or curtained in with hides so that each sleeper had privacy. A screen was sometimes erected between the doorway and the fire, and all the property of the inmates—weapons, household utensils, medicine bags, bullboats, dogs, and even horses—were bestowed inside.

Such lodges were once found all up and down the Missouri and its tributaries, for the earth lodge was the dominant type of dwelling for centuries all over the valley. Such lodges may still be seen on the reservation and also on the State Capitol grounds at Bismarck, North Dakota. They were light, airy, and spacious. When heavy rains set in, the smoke hole could be covered by an overturned bullboat to keep out the wet.

For safety's sake, from both enemies and floods, these lodges were grouped together inside a palisade of vertical logs high on a bluff above the river. In winter the people moved into smaller lodges of a similar type or into tipis erected on lower ground among the trees.

In addition to these giant molehills used as dwellings, there was usually a much larger Medicine Lodge for ceremonies. In this the Mandans celebrated their bloody Okeepa, of which the most striking feature was the voluntary torture, like that endured in the Sun Dance of the Plains tribes. Every man of any standing in those tribes bore the scars of such torture, made as a sacrifice for others, or as a result of a vow made in time of danger, or to ensure success on the next warpath. The artist Catlin made paintings of these bloody rites.

In the Mandan village the ceremonies began about the "Big Canoe" in the center of the plaza. The Big Canoe was a circular enclosure of wooden slabs ten or twelve feet high planted vertically in the ground, the whole being some four feet in diameter. The Indians had a tradition connecting this structure with a great flood from which they had been saved in a big canoe. This legend of the flood went far to convince white men that these Indians were actually descended from the Lost Tribes.

After the drummers had started the music and a mourning woman with hair cut short and bleeding gashes had lamented the loss of some relative killed in war, the male dancers bounded into the open space, all painted grotesquely and wearing masks made of the skin of a bull buffalo's head with painted eyes and real horns, waving green branches of willow. The people perched on the house roofs all about yelled and shouted together, dogs barked excitedly. The dancers became ever more active. Sometimes they butted each other with their heads and, in fact, imitated all the motions of a herd of buffalo. Meanwhile a tall pole was set up, with long cords or lariats dangling loose from its top.

In the midst of the excitement, the drummers suddenly ran into the medicine lodge followed by the young men who were to undergo torture.

Each of these had fasted for several days and drunk no water. He was almost naked and smeared from top to toe with white clay, the costume of a suppliant. Lieutenant H. E. Maynadier has penned the classic account of this ceremony. His journal was published in Washington, 1868, as part of the *Exploration of the Yellowstone River* by General W. F. Raynolds:

"This young man . . . looked ghastly. Kneeling on the ground, one of the old men took up a portion of the skin of the young man's breast and passed a knife through it, making two apertures with a strip of skin between. The blood trickled down, and the victim winced perceptibly. A skewer of wood four inches long was passed through the two holes, and the loop at the end of one of the cords placed over its two ends. The second cord was fastened in like manner to the other breast, and the poor wretch lifted to his feet. The drummers thumped, and the young man threw himself violently back, bearing his whole weight on the cords, and swinging round the foot of the pole. The skin drew out several inches, and seemed to stretch further at every jerk of the poor fellow, who pulled, and tossed, and shouted in order to break away. It was sickening to behold, especially when, after four or five minutes, nature claimed her sway, and the poor wretch fainted and hung collapsed. He was not touched, and seeming to revive, renewed his efforts to bring the torture to a close by breaking the ligaments of skin which held the skewers. After half an hour or more the skin broke, and he was carried off.

"The next victim was served even more dreadfully, though he bore it remarkably well. The skewers were passed under the skin of the back, just above the shoulder blades, and he was hung up to a scaffold with his feet three feet from the ground. Then more skewers were inserted in the fleshy parts of the arms and legs, and buffalo skulls hung to them. I was amazed to see how far the skin would stretch, puffing out to a distance of 12 or 15 inches.

"These disgusting scenes were repeated during two days, varied by races round the big canoe by troops of young men and boys, dragging from four to ten buffalo heads attached to skewers in their backs. Some fainted and did not recover; some were violently nauseated, and proved conclusively that their three days' fast had not been faithfully kept; others held out to the end, and leaped, kicked, and struggled until they were free from their disagreeable attachments.

"All the implements, skewers, bull heads, cords, and willow branches were deposited inside the big canoe, and were considered sacred from that time on. . . . The self-torture and mutilation which accompany their mysteries cannot be explained, except by the supposition that it is a course of preparation for the hardships and dangers of war. I noticed that every male over 10 years old had the scars of the skewer holes on his breast and back."

Of the Mandans, Chief Mato-Tope, or Four Bears, was best-known to the whites. He was a noted warrior and—more important, perhaps, for his fame—an excellent artist who painted his warlike exploits dramatically upon tanned buffalo robes, afterward supplying the necessary explanation. One of these painted robes was given to Catlin, the artist. Prince Maximilian obtained another which—at any rate before the war—was still to be seen in a European museum. Both Catlin and the prince give lengthy accounts of this man.

His most extraordinary exploit was a deed of vengeance for the death of his brother.

One morning the Rees attacked the Mandan village. Four Bears's brother was among those missing. Some days later the chief found the body of his brother shockingly mangled with a handsome spear left sticking through his heart. Four Bears brought the spear back with him into the village, where it was identified as belonging to a Ree warrior of note named Wongatap. Four Bears carried it about, weeping and wailing and vowing that he would revenge the death with that very weapon.

For four years he kept the spear in his lodge, but found no opportunity to use it. Finally, impatient at further delay, he again marched through the village, brandishing the weapon. He declared that the blood from his brother's heart on the blade was still fresh and warned the tribesmen not to mention him nor ask for him nor inquire where he had gone until they heard him shout his war cry at the entrance to the village. "Then," he said, "I will enter it and show you the blood of Wongatap."

Away went Four Bears on his lonely mission of revenge. The Ree village where Wongatap lived was then several days' journey away. Four Bears traveled by night and lay hid in the daytime, living on a little parched corn which he carried, until he reached the town. Apparently, he had already informed himself as to the location of his enemy's lodge. Wrapped in his robe, he boldly entered the enemy town at twilight, mingling with the throng, watching his victim. At last Wongatap went into his lodge.

Four Bears, peering into the lodge in the darkness, waited and watched him light his last pipe and smoke it to the end, saw the last curl of blue smoke faintly floating from its bowl. Finally he saw his victim crawl into bed with his wife.

Waiting until the whole village sank into darkness and silence, until the fire in the middle of the lodge had dwindled to embers, he walked softly, but not slyly, into the home of his enemy and seated himself by the fire.

A large iron pot hung there with a quantity of boiled meat and soup in it. Deliberately Four Bears turned to the pot, took up the big spoon of yellow ram's horn, dipped up what he wanted, and coolly made a hearty supper. Afterward, just as deliberately, he loaded and lighted Wongatap's pipe and made burnt offering to his gods, no doubt with a strong prayer for victory.

Wongatap had gone to sleep, but his wife, lying beside him, several times nudged him in the ribs and asked, "Who is that eating in our lodge?"

Wongatap, being a famous man, always kept open house to all comers. He considered it beneath his dignity to ask anyone what he was doing there. Any man in the village was free to walk in and help himself. He therefore replied, "It's no matter, let him eat. Perhaps he is hungry."

Again while Four Bears was smoking the wife waked her husband and whispered, "Who is that man smoking by the fire?"

But again Wongatap replied, with some impatience, "Let him smoke. Maybe he has no tobacco."

And so, at last, the woman slept also.

Four Bears, of course, had counted on this. He was not alarmed. Now, leaning back and turning gradually on his side to get a better view of his enemy, he stirred the embers with his toe. A little flame sprang up. Then, seeing his way clearly, Four Bears seized the lance, jumped up and plunged it through the body of his enemy. Quickly snatching the scalp from the head, he darted from the lodge, lance in one hand, scalp in the other, and made his way to the open prairie.

By that time the village was in an uproar. All night Four Bears ran. He lay hid during the following days, praying fervently that he might get home in safety.

On the sixth morning at sunrise, he reached home and entered the village amid the war cries of the men and the high-pitched ululations of the women, triumphantly brandishing the spear now all freshly covered with blood and adorned with the scalp of Wongatap. . . .

But the glory of the Mandans was soon to end. In 1837 smallpox came upriver on a steamboat. The traders, fearing an epidemic and the loss of all their customers, sent messages of warning to all the tribes to keep away from the fort. But the Indians, cocky and self-confident, laughed and ignored the warning. They came piling in as usual and camped around the stockade. The traders locked the gates against them, telling them to go away and save themselves alive. They even stood on the roof and held up to view a child covered with smallpox scabs to show the danger. But the Indians had no ears.

The plague swept their villages and spread far up the Missouri and over the surrounding plains to other tribes, destroying at the very least ten thousand victims—and, by the estimate of Kenneth McKenzie, two or three times that number.

The Rees and Hidatsa lost at least half their people; the Mandans were reduced to thirty families.

Smallpox had visited them before, but never with such fury. Some attempted to cure the disease (as they cured others) by taking a sweat bath. Those who did not die in the steam afterward staggered out to plunge into the river and drowned there.

Schoolcraft quotes an eye witness as follows:

" 'Language, however forcible, can convey but a faint idea of the scene of desolation which the country now presents. In whatever direction you turn, nothing but sad wrecks of mortality meet the eye; lodges standing on every hill, but not a streak of smoke rising from them. Not a sound can be heard to break the awful stillness, save the ominous croak of ravens and the mournful howl of wolves fattening on the human carcasses that lie strewed around. It seems as if the very genius of desolation had stalked through the prairies and wreaked his vengeance on everything bearing the shape of humanity.' "

Another writes: "Many of the handsome Arickarees, who had recovered, seeing the disfiguration of their features, committed suicide; some by throwing themselves from rocks, others by stabbing and shooting. The prairie has become a graveyard; its wild flowers bloom over the sepulchres of Indians.

" 'The atmosphere, for miles, is poisoned by the stench of the hundreds of carcasses unburied. The women and children are wandering in groups, without food, or howling over the dead. The men are flying in every direction. The proud, warlike, and noble-looking Blackfeet are no more. Their deserted lodges are seen on every hill. No sound but the raven's croak or the wolf's howl breaks the solemn stillness. The scene of desolation is appalling beyond the power of imagination to conceive.' "

By scattering over the prairie in all directions, the Sioux saved themselves, and the village tribes, now at the mercy of these enemies, moved upriver to Fort Berthold and combined forces in one town. Today their descendants live together on the same small reserve and are much intermarried. Nevertheless, they maintain their own separate tribal organizations and customs and languages. Their wars are over. But they have not forgotten their former glory, and still entertain a healthy intertribal jealousy.

When the bridge across the Missouri was built, the Mandans proposed that it be called after Four Bears, their famous chief, whom Catlin and Bodmer painted and about whom so much had been published.

The Hidatsa also wished the bridge named Four Bears Bridge after one of their chiefs who died just before the bridge was built. The white men, in order to settle the dispute, put up a plaque at the western end of the bridge honoring the Mandan chief, and another plaque at the north end honoring the Hidatsa chief.

By this time the Arikara had decided to sit in on the argument; they demanded a plaque honoring several chieftains of their own.

Immediately the other two tribes nominated a number of chiefs whose relatives felt that they should also be honored.

Finally the feud was settled by putting up a plaque at each end of the bridge bearing the names of the following chieftains—*For the Mandans:* Charging Eagle, Red Buffalo Cow, Flying Eagle, Black Eagle, and Water Chief. *For the Hidatsa:* Poor Wolf, Porcupine, Crow Paunch, Big Brave, Crow Flies High, Big Hawk, and Old Dog. *For the Arikara:* Bear Chief, Son of the Star, White Shield, Bob-tail Bull, and Peter Beauchamp. Nineteen in all!

But as nobody can possibly remember all these names or prefix them to the word "Bridge" in ordinary talk, the structure is now known to everyone as the Four Bears Bridge.

These tribes cannot even agree upon a name for the Missouri River. The Hidatsa call it Anati, which has been translated "navigable stream full of dirt." The Mandans call the river Mata, which may be translated as a "boundary between two pieces of land." Thus the one tribe thinks of the stream as a *highway* and the other as a *boundary*. Each is right.

Nearly every tribe on the Missouri had some interesting skill that set them off from other Indians. Thus the Mandans knew how to make glass beads, beautiful thin pottery, and were famous for their artistic skill in painting warlike exploits upon tanned buffalo robes. The Osages had a class of men who served as chefs or cooks, devoting themselves to the culinary art, to preparing and presiding over formal feasts, and also acting as town criers. But the Arikara, or Rees, who at one time occupied thirty-two villages up and down the river, were famous for their skill at magical performances and sleight of hand.

These jugglers were medicine men who used these tricks to advertise their powers and inspire awe and confidence in their patients. They held ceremonies lasting for days on end. Their performances were so popular and so frequent that they were known to resident traders as "the opera." Each band had its own special type of sleight-of-hand stunts, according to the kind of disease of which it made a specialty.

Is it possible that these performances, advertising miraculous cures, were the originals of the "medicine show," so familiar to all Americans a generation or two ago? Certain it is that the white man's medicine show usually followed much the same pattern. The program comprised music, dancing, and sleight of hand, and frequently Indians, more or less authentic, took part. Often the panaceas sold to credulous villagers after the free show were advertised as being Indian remedies. Throughout the States, people commonly believed that the red men knew of cures far more effective than those afforded by our own pharmacopeia. This faith must have had some basis, and certainly the performances of the Arikara on the Missouri must have done something to encourage it.

Such medicine shows on the Missouri were usually given at night, when the great domed earth lodge was lighted only by the fire in the center. But sometimes they were performed outdoors in broad daylight.

Thus, doctors who treated burns would demonstrate their powers by taking meat barehanded from a kettle of boiling water, setting something on fire and then eating it, or by walking barefoot over hot stones.

Other magicians would permit themselves to be bound with thongs, foot to foot and finger to finger, enclosed in a net, wrapped in a buffalo robe, and buried in a hole covered by a flat stone. In no time at all the primitive Houdini would throw off his bonds and appear on the edge of the staring throng, free as air.

Some of their tricks are more easily explained. Thus, a dancer would place a gourd rattle under a buffalo robe. Then, stamping upon the robe, he would crush the gourd to bits, lift the robe and show the pieces. Then covering these again for a moment, he would turn the robe back and show the gourd intact.

Another man could pour a bowl of Indian corn on a buffalo robe and "make the robe eat it." Sometimes small dolls would dance alone or smoke a pipe presented to them, or a stuffed bird would whistle or cry in answer to its owner. Another performer might take a live bird or a jack rabbit out of his mouth. Others had "power" to dive into the river and bring up a living fish in either hand.

But the most remarkable stunts were performed by doctors whose business it was to act as surgeons and heal wounds. One of these could swallow an arrow. Or he could stick a skewer through his tongue, take out his knife and cut his tongue off, bleeding profusely. The tongue in his hand would then "turn into" the tongue of a buffalo. This he would put back into his mouth, pull out the skewer, and behold—not a scar would remain to show that he had ever been wounded.

Another medicine man might stick a knife down his throat, turn a back somersault, and "break off the hilt against the ground. Blood would gush copiously from his mouth and he would writhe about the floor apparently in great agony. A fellow medicine man would rush in, pick up the broken hilt, and insert it into the injured man's mouth. The knife would become whole again, and would be withdrawn."[1]

Sometimes a medicine man would allow another to cut off his arm—or even cut it off himself. Then he would pick it up and throw it out among the people, where it appeared to be the leg of a buffalo. When he replaced it, it grew on again as his own arm, leaving no scar.

But the most amazing of all their tricks was that of beheading a naked man. They cut off his head with a saber, and carried it outside to show the people. All the time the bleeding, headless trunk danced gaily around the lodge. Bringing the head back, they put it on the shoulders, but facing backwards. Suddenly, the head appeared in the correct position. The medicine man then rubbed the bleeding cut with his hand. It disappeared, and the decapitated man appeared unscarred, unhurt.

Such tricks imposed not only on Indians but on most of the fur traders, army officers, and boatmen privileged to behold them.

Nowadays it is little realized what a lively night life

Indians in the old days enjoyed. There was a dance or ceremony almost every night, and something going on during every hour of the twenty-four. A big camp never slept.

For the Plains Indians were humorous, jest-loving, merry people in the old days, and their old-timers are so still. In fact, the older they get the merrier they seem to be. Of them all, Gray Whirlwind was the jolliest old buck I ever knew. He was always laughing—not from senility, for he had all his faculties, but out of sheer delight in life and plain good humor.

The last time I saw him he was ninety-six winters old and as merry as a cricket. It was in the summer, at the season when the tipsin is ready for harvest. This white fibrous root or tuber is covered with a brown rough bark, and when dug up looks rather like an oversized mouse. It has to be harvested within a few days' time. For as soon as the plant has flowered, it breaks off at the ground level and the winds blow it away, leaving the root hidden in the earth. So the women go out on the prairie when the plant is flowering and with pointed sticks or hoes dig it up. The children run about over the prairie, finding the plants for their mothers to dig.

I found my old friend far out on the grass out of sight of any house, sitting alone in his tent while his children and grandchildren and great-grandchildren were busy at the harvest all about.

He made me welcome. We talked on until midday.

The flap of the tent was open, and I could see for miles across the gray-green Dakota prairie. The tent, pitched only a few hours before, inspired a carefree feeling, since the grass inside it had not yet been trodden down. The very weather seemed to agree with the old man's jollity.

While we sat there talking of the old days, a furry half-grown puppy came bouncing in and flung himself upon me. He was sure the whole world was his friend, and the rest of the morning he was in my lap or on my feet every moment. I had to pet him to keep him from breaking up the meeting.

At noon my interpreter turned to me and said, "The chief wishes to honor you. He wants you to stay for dinner."

Of course I was pleased at that, more particularly as it was many miles to a restaurant and I wished to continue my talk.

"Tell the chief," I said, "that I shall be very glad to accept his hospitality."

The interpreter hesitated. Instead of addressing Gray Whirlwind, he spoke to me again. "He is going to feed you that puppy," he said, eying me expectantly.

Now I knew that to be invited to a dog-feast among the Sioux was like being invited to a turkey dinner by a white man, and I knew too that dog meat has always enjoyed the reputation of being the finest flesh available on the prairie. I appreciated the honor done me. But I said to the interpreter, "Tell the chief we have to hurry back to town. I might eat dog and like it, but I'm damned if I will eat a dog I have just been playing with!"

CHAPTER XIX

Custer and Comanche

FORT ABRAHAM LINCOLN stood on the west bank of the Missouri just below the mouth of Heart River. As at other frontier posts, the buildings were arranged in an open square about the level parade ground and were protected by a low range of bare ridges on the landward side. That fort is forever associated with George Armstrong Custer, "the boy general," "the flower of cavalrymen"—one of the most dashing victors of the Civil War, who could brag that he had never lost a gun or a flag. Custer generally wore his yellow hair long like a mountain man; he went in for fancy uniforms and fringed buckskin jackets. After successfully attacking the Southern Plains tribes at the battle of the Washita in 1868, Colonel Custer [1] was regarded as the best Indian fighter in the army. He had covered his regiment, the Seventh Cavalry, with glory.

In 1874 Custer was sent from Fort Abraham Lincoln at the head of his regiment to reconnoiter the Black Hills, the sacred mountains of the Sioux. There the Army planned, in flagrant violation of the existing Indian treaty, to build a post to overawe the Indians. The expedition was camouflaged as an exploration and scientific survey. Custer, who had ambitions to be president, penned glowing accounts of that beautiful region, its lovely flowers and rippling streams, wild fruits, rich forage, and sheltered valleys. He also announced to the newspapers that he had found "gold in the grass roots."

GENERAL GEORGE CUSTER

His story started a gold rush up the Missouri River and across the plains to the tune of "Ho! For the Black Hills." The Army made a pretense of keeping gold seekers off the Indian hunting grounds, but in no time at all there were thousands of miners in the Hills. This unwarranted invasion roused the Sioux to fury. The Black Hills were the very heart of their country.

Somewhat later Custer was rash enough to call the attention of the public to graft at the Indian agencies, and went to Washington to testify in the impeachment trial of President Grant's secretary of war, William K. Belknap.

Though personally honest, the president was bitterly offended, particularly as Custer's testimony turned out to be mere hearsay. So when, in 1876, troops were ordered to attack Sitting Bull's Sioux, Grant forbade Custer to accompany the expedition.

It seemed to Colonel Custer that this was the end of his career. Frantically he appealed to Sheridan and Terry, pulling all the wires he could lay hands on. Luckily Terry was anxious to have the services of such a redoubtable Indian fighter, and finally President Grant yielded to his pleas.

Back at Fort Abraham Lincoln, Custer was hell-bent to cover himself with glory in the campaign. He knew that any successful action against Sitting Bull must be fought by cavalry, and he commanded the only cavalry on the expedition. On May 17, 1876, the united command left the fort, heading west to the Yellowstone and the Little Big Horn. Custer had moved heaven and earth to get command of the Seventh. He felt it was almost a miracle that he had it. He was rarin' to go.

The Seventh had only about half its full strength of officers; forty per cent of the enlisted men were raw recruits. In those days there was no serious penalty for desertion, and a great many men had the habit of joining the army at the first snowfall—in order to have shelter and food during the winter months—and then deserting in the spring. Such winter soldiers were known as "snowbirds." Custer himself declared that, had all the men who enlisted in his regiment during that year remained in service, he would have had a brigade.

While the band blared "Garry Owen," the brisk drinking song to which the Seventh marched, the troops moved out with their colors and guidons fluttering gaily in the wind.

The Indian scouts beat their drums and sang their melancholy war songs while their squaws crouched on the ground with hanging heads. Near Laundress Row, the wives and children of the troopers lined the road, and mothers with streaming eyes held their children high to take a last look at their departing fathers. Meanwhile the older youngsters, with handkerchiefs tied to sticks to serve as flags, and beating old tin pans for drums, stepped out in imitation of the soldiers.

The officers' ladies watched from the galleries of their quarters, bravely concealing their distress. But when the mounted band struck up "The Girl I Left Behind Me," wives and sweethearts suddenly burst into tears and fled indoors.

Fate staged this parting in a manner suited to the occasion and to the undying fame to which these men were riding. In her book *Boots and Saddles,* Mrs. Custer has described the scene:

"From the hour of breaking camp, before the sun was up, a mist had enveloped everything. Soon the bright sun began to penetrate this veil and dispel the haze, and a scene of wonder and beauty appeared. The cavalry and infantry in the order named, the scouts, pack-mules, and artillery, and behind all the long line of white-covered wagons, made a column altogether some two miles in length. As the sun broke through the mist a mirage appeared, which took up about half of the line of cavalry, and thenceforth for a little distance it marched, equally plain to the sight on the earth and in the sky." [2]

So Custer and his doomed men rode away into the empty sky.

The colonel, knowing that his route lay through the Badlands and that speed was essential to victory, refused to take along the Gatling guns. In order to avoid clatter on the march, all sabers were left at the fort.

The whole command moved to the Yellowstone. There they divided into three forces: Terry led one, Gibbon the other, and Custer his beloved cavalry. They planned to meet on the Little Big Horn and crush Sitting Bull's camp by a pincer movement.

Terry, the commanding general, was new to Indian fighting, and therefore did not pin the experienced Custer down with precise orders. When the Seventh rode off, while the trumpeters blasted the air, Custer halted to salute his commander. One of the officers called out, "Now, Custer, don't be greedy; wait for us."

Custer grinned and yelled back, "No, I won't." Nobody has ever been quite sure what he meant by that.

Custer's orders were that, if he found a big Indian trail, he was to send word to Terry. But instead he pushed on hard after the Indians, in spite of warnings from his scouts that he would find more Sioux than he could handle. There were unlucky omens, too. The flag before his tent twice fell down, and at Officers' Call before the fight men noticed a curious despondency in Custer quite alien to his usual buoyancy. His long hair was missing, too; he had got a haircut.

For three days Custer rode hard on that Indian trail. It seemed his great opportunity. He was keen to redeem himself, and was never the man to refuse a fight. Fearing that the Indians might learn he was coming and run away, he neglected to make adequate reconnoissance, divided his small force: one part under Benteen, one under Reno. He himself took five troops, the bulk of the command, and disappeared behind the bluff waving his hat.

Benteen had to go roundabout and so Reno, green at Indian fighting, struck the Indian camp first of all. The Sioux fought hard. Within a short time, Reno ordered a retreat which soon became a rout, and scrambled up the bluffs across the river where he dug in to save what was left of his command. There Benteen joined him. Their combined force was hemmed in by the Sioux and held inactive. Meanwhile, three miles downstream on the steep broken ridges, thousands of Sioux, Cheyennes, and Arapahoes swarmed over Custer and his troopers, wiping them out to the last man.

Custer might have taken warning from the fate of two other commanders previously wiped out by these same Indians. They had exterminated a detachment at Fort Laramie in 1854 in the Grattan affair, and again at Fort Phil Kearny in 1866 when they rubbed out Captain W. J. Fetterman and eighty-two men in forty minutes. But now, on the Little Big Horn, they scored an even greater triumph.

Next day, after fighting with Reno on the bluffs, the Indians pulled out for the Big Horn Mountains. Terry and Gibbon came up the river and found the bodies of Custer and his men. Custer fell on June 25th. When the Crow scouts brought the bad news to Captain Grant Marsh on his steamboat the *Far West*, Marsh forced the boat up to a point only fifteen miles from the battlefield. There Reno's wounded were carried aboard.

Marsh turned his boat around, and by a miracle of navigation brought it safely down the Big Horn, down the Yellowstone, down the Missouri, and on the morning of July 5th made landing at the fort. There he was met by twenty-six weeping widows. He first broke the news of this terrible disaster to the American public.

Everything seemed to conspire to make Custer's last stand a landmark in the annals of the West. Fate had arranged that it should occur when the whole country was celebrating the centennial of the founding of our nation.

The steamboat also brought one lone survivor, badly wounded, down from the battlefield after the fight. This was a horse, Comanche, which had been the mount of Lieutenant Keogh. The horse was carried in a wagon to the fort, and for a whole year rested in a sling in his stall and was served with a frequent whisky-bran mash. When he was able to walk again, he became the fetish of the Seventh Cavalry.

A general order was issued that Comanche should never be ridden or put to any kind of work again. At every parade of the Seventh Cavalry, this horse, saddled and bridled, was led by a mounted trooper. Sometimes it would break away and lumber up to its old place at the head of the troop.

The Seventh Cavalry observed the anniversary of Custer's last stand as a day of mourning and usually held a ceremonial parade. Then Comanche marched before "I" troop, draped in a black mourning net, with saddle reversed and riding boots reversed in the stirrups.

During the rest of the year he was free to do pretty much as he liked. His stall had no gate, and he wandered about the post in perfect freedom. No one dared object to his doings, since he was regarded as "the second commanding officer" of the regiment. He grazed freely on lawns and flower beds, upset garbage pails, and begged sugar from door to door. He had an ear for music, too, according to his biographer, Captain Edward S. Luce,³ as he always turned up on summer evenings for the regimental band concert and grazed around the bandstand.

He was cared for by Custer's personal orderly, Private Burkman, affectionately known to Custer as "Old Nutriment," and by the regimental blacksmith. On payday the men regaled the horse with buckets of beer, and Comanche throve on it, dying at last at a ripe old age in 1891. Afterward the horse was mounted by a taxidermist, Professor Lewis Lindsay Dyche of the University of Kansas, and may still be seen, saddled and bridled as of old.

But by the time Comanche died, the Seventh had had its revenge.

CHAPTER XX

Ghost Dance

After the Custer "massacre" public feeling throughout the country ran high against the Indians in the West—particularly the Sioux. Pony soldiers (cavalry), wagon soldiers (artillery), and "walk-a-heaps" (infantry) swarmed into the Missouri Valley and soon chased Sitting Bull and his hostiles over the line into Canada. There his Sioux remained as long as the buffalo lasted, dribbling back at last in small bands to surrender at one or another of the Missouri posts until Sitting Bull handed over his own gun at Fort Buford at the mouth of Yellowstone in July, 1881. He was sent downriver on a steamboat and held prisoner at Fort Randall for two years; then he was permitted to join his tribe, the Hunkpapa, on the Standing Rock Reservation along Grand River.

During the early eighties there was undeclared war in the country between the Rockies and the Big Muddy. All that region was being rapidly transformed, as white hunters swept away the buffalo and other game. The railroads brought in settlers, and ranches, farms, and towns sprang up everywhere. Many of the white men in that country regarded the Indians as fair prey, stealing their horses or shooting them down whenever they could do so with impunity.

The Indians, on their part, found themselves defeated, stripped of their guns and ponies; cheated by thieving agents, grafting contractors, and governmental red tape of the rations and annuities guaranteed them by the treaties; submerged in a hostile population, and starving. Their religious ceremonies were suppressed, their political organizations shattered, their leaders murdered, and their children were taken away as hostages to distant schools or destroyed by strange epidemic diseases of the white men. And all about them they felt pressure forcing them to cede their reservations for a song.

The new treaties too often were railroaded through without the consent of the tribes whose "chiefs" were bribed or forced to sign. Their reservations were not always surveyed. The very earth seemed sliding from beneath their feet. They saw themselves being liquidated. The white man's road appeared to lead only to the Indian's grave. Their very gods were overthrown.

Having no guns, few ponies, and being unable to subsist on the meager rations supplied them at the agency, they made bows and arrows and scoured the country—in which by treaty they were permitted to hunt—off the reservations. As the game disappeared, they sometimes made free with the white man's beef. It was a simple matter for an Indian to kill a steer silently with his bow in a coulee while his comrade kept watch on the ridge, ready to signal to him with a sunflash from a hand mirror in case a cowboy rode into view. The white man had exterminated the Indian's buffalo; the Indian was killing the white man's cattle. Turn about is fair play.

Steers sold to the government for Indian subsistence were usually bought in the fall when they were fat and in prime condition, yet might be issued in the spring when they were little more than horns, skin, and bones, Sometimes the weight of such an animal was reduced by one-half during a northern winter.

Summer after summer the Indians were held in council when they should have been harvesting their scanty crops, and by the end of the eighties severe drought had made farming in that country impossible. White settlers pulled out to the East by thousands, and the population of Bismarck, the capital city of North Dakota, dropped to five hundred souls. . . .

In 1889 the Sioux first heard of the Indian Messiah. The news was heartening.

Despairing of their own gods, they turned now, ignorantly to worship the unknown God of the white man.

The Messiah, a Pai-Ute Indian named Wovoka, lived far away in Nevada. Some years before, during an eclipse of the sun, he had fallen into a trance and had a revelation. The story went that he was the Saviour born on earth again—this time in the flesh of an Indian, because the white men had crucified him at his first coming—that he would come from the west preceded by a wave of earth which would cover the whites and all Indian unbelievers, and followed by vast herds of buffalo, horses, and all the Indian dead brought miraculously to life again. The Indians who believed in him were to inherit this new earth and live forever in the happy hunting grounds. Everything was to be as it had been before.

To share in this millennium his followers had only to dance the dance he taught them, holding hands in a circle and moving to the left four nights in a row at intervals of six weeks.

Wovoka was insistent that no arms were to be carried in the dance. His religion was one of peace. He told his followers that they must not be afraid at the earthquake which would precede his coming—the feathers they wore in their hair would waft them above the landslide and save them alive.

His moral code was simple: "Do no harm to anyone. Do right always. Do not tell lies. Do not fight. When your friends die, you must not cry."

The Sioux promptly sent delegations to investigate the new doctrine and see this Messiah. They were gone all winter, and on their return in the spring of 1890 reported that the Messiah would come in the following spring to redeem his people. The result was a tremendous religious revival. Many of the Indians abandoned everything the white men had taught them, and went zealously to work to carry out the instructions of the Messiah.

It was reported that the Messiah had the marks of the nails in his hands and feet, and apparently he was something of a sleight-of-hand artist, and perhaps a hypnotist. When one of the delegates looked into Wovoka's upturned hat, he "saw the whole world."

In the dance the leaders threw the more susceptible worshipers into trances, waving feathers above their eyes as had formerly been done in the Sun Dance. In these trances visions were seen, and these visions, naturally enough, confirmed the doctrines taught in the ceremony. People met and talked with their dead relatives and sometimes brought back trinkets or buffalo meat from the spirit land.

The cult spread like wildfire and affected nearly all the Indians west of the Missouri. Naturally those tribes who were most in need and most unhappy were the most zealous believers. On reservations all up and down the river, the new doctrine found believers and dances were held at Rosebud and Pine Ridge, at Cheyenne River and at Standing Rock, Fort Berthold, Fort Peck, and Fort Belknap. Kicking Bear brought the dance to Sitting Bull's camp on Grand River in October, 1890.

The Sioux well remembered how their Sun Dance had been suppressed, their medicine men forbidden to practice, and all their rituals frowned on by their agents. They had gone through a terrible ten years of war and famine, and they had forebodings that the white men would again try to put a stop to their dancing. It was not long until they had evidence enough of this.

Warlike as they were, they had long been accustomed

to protecting themselves in battle by painting religious symbols upon their shields, their bodies, their clothing, and their horses. In their visions of their dead, they naturally saw such painted garments, which had made their fathers proof against bullets and arrows. So now, fearing that even the women and children would be massacred, they all wore shirts and dresses made of old flour sacks or buckskin and painted with the sun and moon, morning star, the buffalo, turtle, crow, and eagle. These garments, known as "ghosts shirts" among the whites, were worn by all alike—men, women and children—in the dance. One of their leaders told them, "The guns of the soldiers are the only things we are afraid of. But these belong to our Father. He will see that they do us no harm."

Covered with this spiritual armor, the Indians went ahead with their dance. It seemed impossible to stop it, any more than one might hope to stop a religious revival among the more enthusiastic sects among the whites.

Unhappily, when in 1888 the national Republican landslide had ousted President Cleveland and the Democrats, many experienced Indian agents were replaced by green men. Such political appointees, understanding nothing of agency factions, naturally applied the spoils system on their reservations—on the absurd theory that all the Indians, half-breeds, and white men who had sustained the former Democratic agent must be replaced by members of the opposing, and inevitably reactionary, group. Thus at Pine Ridge, the new agent, D. F. ("Young-Man-Afraid-of-the-Indians") Royer in one day destroyed the progressive organization which his predecessor, Dr. V. T. McGillycuddy, had built up throughout long years of service. McGillycuddy was a genius at handling Indians and had never required any troops to back him up. He had put trust in those Indians willing to work for the advancement of their people.

Yet the Pine Ridge Reservation was probably the toughest assignment in the Indian Service. It was very large and swarmed with thousands of unreconstructed Sioux led by that old fraud and mischiefmaker, Chief Red Cloud. And now Royer, the new agent, wholly ignorant of Indians and their ways of thinking, found himself at the mercy of Red Cloud's faction of inveterate troublemakers. For the progressive Indians, finding that the new agent would have none of them, let him go to the devil in his own way.

It was not long before Royer was yelping for troops to protect him from his bloodthirsty "Republicans." Unfortunately General John R. Brooke, who was sent to suppress the "uprising" was little better qualified. Neither of tthe men showed any foresight, any understanding. The attitude toward the Indians of the agent was *"Please be good"*; of the general, "I will pacify them."

The white people who lived on the Missouri River were slow to take alarm; and as winter was coming on most of them thought the dance would die away out of sheer bad weather. But the newspapers of the country at large foresaw another Indian campaign and sent their correspondents, armed to the teeth, swarming out to the Missouri to be in at the death. Few wars in American history have been so thoroughly covered as the so-called Ghost Dance War.

Unfortunately, the reporters did not contact the only men who could have told them what was really going on—the Indians.

Here are some opinions of prominent Indians, most of them of the progressive faction:

Young-Man-Afraid-of-His-Horses: "The Indians are brave and the white men are brave, but the white men do not do as they agree; that is the trouble with the Indians."

Two Strike: "My heart is good. I am for peace; I am not for fighting, but *we had rather die fighting than be disarmed and then killed.*"

Kicking Bear: ". . . When the Agent robs us they send soldiers."

Little Chief: "My people are warriors. If the Government does with us as it agreed, they are peaceful. The Government took away our good land, promised us money and plenty to eat; they said they would bring us to a good country and teach our people to farm and be like white men. They brought us to this country where nothing grows. The agent steals our beef. My people get poorer every day, and when they starve their hearts are sore."

Rocky Bear: "Treaties are only a lot of lies. The Government never kept any treaty it ever made with us. We have always been robbed and lied to. We did not commence the fight."

Crow Dog: "We want what was promised. We want to do right, but we do not want to have our guns taken away and be treated as slaves!"

American Horse: "The Indians and the whites should all live as one."

Big Road: "When I promise to do something, I do it. When the Great Father promises he never does it." [1]

When the newshawks arrived, they found little enough to report, and not a few of them, under pressure from the home office to send in thrilling stories, did a nobler thing than record history—they made it up.

Even when their stories were not false, the headlines placed upon them by the editors were often sensational. And the correspondents sent in so much copy that they had to make up a schedule at the telegraph office so that everybody could wire his own story back every twenty-four hours. Each day the newshawk at the top of the list dropped back to the bottom. They kept the wires hot.

The troops arrived on November 19th, and their coming precipitated an immediate stampede of frightened Indians from Pine Ridge into the Badlands to the West. At that time the United States Census shows that whites outnumbered Indians in Dakota seventeen to one!

BUFFALO BILL

Having run the Indians out, the army now prepared to bring them back—at first by negotiation and afterward by sterner methods. Meanwhile, at Standing Rock, Major James McLaughlin was doing his best to prevent a similar scare on his own reservation. He declared that there would be no trouble with the Indians unless it were forced upon them, but took advantage of the general alarm to recommend the arrest of Sitting Bull. For "Old Bull" still claimed to be head chief of his people and was a thorn in McLaughlin's flesh.

The Indian Bureau paid no heed to McLaughlin's suggestion, since the agent could not offer any valid reason why Sitting Bull should be arrested.

About that time Buffalo Bill, finding that the hero business in which he was engaged was slumping, saw an opportunity to get some free publicity and increase the income of his Wild West Show. At a dinner he persuaded General Nelson A. Miles to scribble him an order for Sitting Bull's arrest.

Cody hastened out to Standing Rock and displayed his authority to make the arrest. McLaughlin and his friends, the officers at Fort Yates nearby, were unwilling to have an outsider come barging into their preserves at such a critical moment. Moreover, if an arrest was to be made, they wanted to make it themselves.

Accordingly, the officers conspired to delay Cody while McLaughlin busily wired to Washington, trying to get the order rescinded. The officers all knew that Cody was a convivial soul, so all one night they entertained him in relays at their club, trying to drink the famous showman under the table.

Such an attempt was like trying to fill an Indian up with beef—doomed to failure. Next morning, bright as ever, Cody set out for Sitting Bull's camp to bring the old man in. Sitting Bull had served in Buffalo Bill's show throughout the season of 1885, and Cody had no fear of him.

Desperate, McLaughlin sent messengers to mislead Cody and delay him until he himself could hear from Washington. And in this the agent was successful. Cody's order was rescinded.

After Cody pulled out in disgust, military orders arrived for the arrest of Sitting Bull. McLaughlin persuaded the army officers to let him send his Indian policemen first "to avert bloodshed." The troops followed somewhat later.

The policemen, some of them personal enemies of Sitting Bull and his friends, or members of the two other Sioux tribes on that reservation knew that they were on a dangerous mission. But they were brave young men. They slipped into his cabin and dragged him naked from his bed before anyone in the camp knew they were about.

It was cold weather, and by the time Sitting Bull was dressed and ready to go, his followers had gathered and surrounded the cabin in the dark, drizzling dawn. The forty policemen found themselves hemmed in by an angry crowd of seasoned warriors who could never endure to let Indians of other bands enter their camp and carry off their chief. Catch-the-Bear shot down Bullhead, chief of police, thus precipitating a fight in which, within a few minutes, Sitting Bull and twelve other Sioux were killed and two others mortally wounded. Had the troops not arrived soon after, perhaps not one policeman would have lived to tell the tale.²

When the troops arrived, the Indians offered no opposition whatever to them, and withdrew without firing a shot. Some of them fled southwest to find Big Foot's camp on the Cheyenne River Reservation.

These Sitting Bull refugees, thirty-eight in number, reached the camp of Big Foot hungry, footsore, almost naked. They were his relatives. He took them in and fed them. As he explained it later, "no one with a heart could have done any less." ³

At the time, Big Foot's band were peaceably encamped on their reservation and about to go to the agency to get their rations. But the killing of Sitting Bull had thrown the fat into the fire; suddenly the military ordered Big Foot's arrest. The army would not let his small band *go* to the agency—it proposed to *take* them there.

Big Foot readily agreed to go. But one of the white scouts talked so threateningly that the other Indians became frightened; that night they lit out for Pine Ridge, traveling light, taking their chief with them.

The troops went after Big Foot and his frightened people, rounded them up without any trouble, and brought them to the Pine Ridge Reservation. There they went into camp on the creek called Wounded Knee. Among the troops in charge of these Indians were several units of Custor's old regiment—the Seventh Cavalry.

The total military force consisted of 470 men, including scouts, with four pieces of light artillery, as against 106 warriors with Big Foot. That morning the chief himself was in bed with pneumonia. It was December 29th, 1890, just two weeks after Sitting Bull was killed.

The Sioux tipis stood west of the creek along a dry ravine running into it. Not far away, on a slight elevation, four Hotchkiss guns were posted, *trained on the Indian tents*.

Troops were stationed on all sides of the camp. Colonel James W. Forsyth was in command. He had orders from General Brooke to disarm the warriors. The general wished to take them to the agency and on to the railroad.

The interpreter told the warriors to come out of their tents. They came out wrapped in their blankets and sat on the ground facing the troops. They were then told to go into the tents and bring out their weapons. It was against all Indian custom to surrender one's arms. Perhaps too, in the face of this humiliation, some of them were slow in order to display their courage. At any rate, only twenty of them went into the tents. They brought out only two guns.

These Indians had had no intention of fighting. There was never any trouble on their reservation. They had made no attacks upon settlements or settlers. They were guilty of no raids or depredations. But they *knew* how, after the Cheyennes had surrendered in Oklahoma, many of the warriors—innocent and guilty alike—had been ironed, imprisoned, sent to Florida, and held for long years away from their families in that hot country. Big Foot's warriors thought they too were going to be so disarmed, imprisoned, exiled. Why else had orders been given to take them to the railroad?

It all seemed very unjust to them. They had done nothing whatever, but had merely been robbed of their rations and were suffering for the necessaries of life. If they resisted, they knew they would be killed, with their wives and children. But maybe it was "better to die all together" than to be separated and starved one by one.

The officers in charge seem not to have had much understanding of Indian ways of thought. They made a series of mistakes.

First, they ordered a detachment of troops forward to within ten yards of the seated warriors. The troops were so placed that, if they fired, they would inevitably shoot each other.

Second, they ordered another detachment to search the tipis.

The search party, of course, could not speak Sioux. They went into the tents, drove out the women and children, and overturned the beds, looking for weapons. The warriors, still armed, could see and hear their frightened wives and children running about, crying with alarm and anger. They knew how often Indian women and children had been abused or killed by white men. It was hard to sit still and wonder what was going on in those tents.

Meanwhile Yellow Bird, said to have been a medicine man, began to walk up and down along the line of warriors, blowing his eagle-bone whistle. He harangued the warriors. Maybe he was just showing off to let everyone know he was not intimidated, as frequently happened when warriors surrendered. Maybe he told them that the soldiers were weak and powerless, that the ghost shirts would turn their bullets. Maybe he urged them to resist.

If the officers had known Indians well, they would have known that when a warrior walks up and down he is nerving himself for a fight.

When the search party left the tents, they brought only forty guns, most of them antiques and of little value. It was clear that, if the warriors had any guns, they had them under their blankets.

Wild Indians were not given to scuffling, fisticuffs, or any such friendly forms of fighting. When a man laid hands on a wild Indian, that Indian assumed that his life was threatened and knew that his personal dignity had been outraged. Unless an arrant coward, the warrior immediately struck to kill. One could not take such liberties with a wild Indian.

One of the soldiers took hold of the blanket of a warrior and lifted it to see if he was armed. In the Indian view this amounted to an attack.

Immediately, they say, Yellow Bird stooped down, scooped up a double handful of dust, tossed it high into the air, like an angry buffalo bull pawing the earth. The chill wind made a broad banner of the tawny dust. It was the old signal for war.

That is the popular white man's version. But Indian witnesses agree with the simpler account given by General Nelson A. Miles that "a scuffle occurred between one warrior who had a rifle in his hand and two soldiers. The rifle was discharged." [4]

Then everything happened at once!

All the warriors jumped to their feet. Immediately the soldiers fired a volley, laying half of them low. The men were standing so near that their guns almost touched. All joined in a fierce hand-to-hand struggle, both sides standing their ground. Some of the Indians had repeating rifles. The troops carried single-shot Springfields. Powder smoke and dust clouded the savage melee. In the blind excitement of that fierce infighting men shot and knifed and clubbed each other, not knowing whom they killed.

The remaining Sioux charged through the line of troopers to take cover in the ravine. Meanwhile the battery on the rise went into action, quickly silencing the camp, killing Big Foot in his bed. Then the guns blasted Indians who were running for their ponies. After that, one of the pieces was limbered up and hurried to the dry ravine to clean out the last nest of resistance.

By that time the surviving Indians were running away down the ravine under a storm of two-pound explosive shells. The artillery mowed them down. Within a short time some two hundred Indian men, women, and children, and sixty soldiers, lay dead and wounded on the ground. The tipis, knocked down and burning, smoked into the winter sky, while the soldiers, in a frenzy of revenge, raced after the flying survivors.

Many of the women and children, some of them infants in arms, were found dead a long way from the scene of action—some at a distance of two miles.

Of course, this massacre was not intended by the officers engaged. In fact, they had separated the women and children from the men to prevent any such disaster. However, many of the troops were recruits who had never been under fire and probably could not, in the dust and smoke and excitement, distinguish an Indian woman from a man. In those days all Indians wore their hair long. The Indians, too, made mistakes. Father Craft, a Catholic priest, or "Black Robe," was wearing a soldier overcoat and a fur cap that day. One of the Indians stabbed him.

By the irony of fate, most of the reporters, having decided that there never would be any trouble after all their yelling "wolf, wolf," had not taken the trouble to come out from the agency and witness the disarming of the "hostiles." Only three newshawks turned up. But one of these, named Kelley, is said to have killed three Indians in the fight. He scooped the other reporters, too, for by good luck that was the very day his dispatches were scheduled to go first over the wire.

This bloody affair at Wounded Knee caused repeated investigations and brought on a fit of buck-passing between the Indian Bureau and the War Department at Washington. It was claimed that Wounded Knee was a "slaughter and without provocation"; that "the military are to blame for all this trouble and that the presidential aspirations of General Miles and the desire of the Seventh Cavalry to avenge Custer's death precipitated the fight at Wounded Knee." [5]

After the fight Colonel Forsyth, with eight troops of the Seventh Cavalry, made a wholly unnecessary dash to rescue the white people at the Catholic mission, who were never in any danger from their Indian friends. But as it turned out, ingloriously, the colonel's detachment had itself to be rescued by Major Guy V. Henry and his "buffalo soldiers." The Indians called the Negro troopers buffalo soldiers because they had wool like buffalo. Thus the Ninth saved the Seventh that day and so put an end to the last Indian uprising on the plains. All was quiet along the Missouri.

It is interesting to note that, though so many newspapermen were sent to cover the Ghost Dance War, the only piece of any sincerity or literary quality written about it at the time is a ballad by a Negro trooper, Private W. H. Prather, of the Ninth Cavalry:

THE INDIAN GHOST DANCE AND WAR

The Red Skins left their Agency, the Soldiers left their Post,
All on the strength of an Indian tale about Messiah's ghost
Got up by savage chieftains to lead their tribes astray;
But Uncle Sam wouldn't have it so, for he ain't bilt that way.
They swore that this Messiah came to them in visions sleep,
And promised to restore their game and Buffalos a heap,
So they must start a big ghost dance, then all would join their band,
And maybe so we lead the way into the great Bad Land.

Chorus:

They claimed the shirt Messiah gave, no bullet could go through,
But when the Soldiers fired at them they saw this was not true.
The Medicine man supplied them with their great Messiah's grace,
And he, too, pulled his freight and swore the 7th hard to face.

About their tents the Soldiers stood, awaiting one and all,
That they might hear the trumpet clear when sounding General call
Or Boots and Saddles in a rush, that each and every man
Might mount in haste, ride soon and fast to stop this devilish band.

But Generals great like Miles and Brooke don't do things up that way,
For they know an Indian like a book, and let him have his sway
Until they think him far enough and then to John they'll say,
"You had better stop your fooling or we'll bring our guns to play."

Chorus:—They claimed the shirt, etc.

The 9th marched out with splendid cheer the Bad Lands to explo'e—
With Col. Henry at their head they never fear the foe;
So on they rode from Xmas eve 'till dawn of Xmas day;
The Red Skins heard the 9th was near and fled in great dismay;
The 7th is of courage bold both officers and men,
But bad luck seems to follow them and twice has took them in;
They came in contact with Big Foot's warriors in their fierce might
This chief made sure he had a chance of vantage in the fight.

Chorus:—They claimed the shirt, etc.

A fight took place, 'twas hand to hand, unwarned by trumpet call,
While the Sioux were dropping man by man—the 7th killed them all,
And to that regiment be said "Ye noble braves, well done,
Although you lost some gallant men a glorious fight you've won."
The 8th was there, the sixth rode miles to swell that great command
And waited orders night and day to round up Short Bull's band.
The Infantry marched up in mass the Cavalry's support,
And while the latter rounded up, the former held the fort.

Chorus:—They claimed the shirt, etc.

E battery of the 1st stood by and did their duty well,
For every time the Hotchkiss barked they say a hostile fell.

Some Indian soldiers chipped in too and helped to quell the fray,
And now the campaign's ended and the soldiers marched away.
So all have done their share, you see, whether it was thick or thin,
And all helped break the ghost dance up and drive the hostiles in.
The settlers in that region now can breathe with better grace;
They only ask and pray to God to make John hold his base.

Chorus: They claimed the shirt Messiah gave, no bullet could go through,
But when the Soldiers fired at them they saw this was not true.
The Medicine man supplied them with their great Messiah's grace,
And he, too, pulled his freight and swore the 7th hard to face.[6]

CHAPTER XXI

The Missouri River Women

WHAT song the Sirens sang to Odysseus is a puzzling question if not beyond all conjecture. But we know that the song sung by the sirens of the Big Muddy. For the Missouri River has its sirens too, who are in fact organized into a society and hold regular meetings in the fall or spring. Some of them are spirits, others good, solid, respectable women of the Mandan village.

Like the ancient Greeks, the Village tribes believe that spirit women, or—so to speak—hamadryads, dwell in the trees and buttes along the river. These strange beings are thought to assemble under a bare peak on the west bank known as Eagle's Nose or Bird Beak Peak. The Mandans were quite sure of this even before they became members of the society.

Once upon a time these spirit women lured away two Mandan women and carried them to their subterranean lodge under the peak. The Mandans mourned the two as dead, not knowing what had become of them, until—after a time—one of the women turned up again and rejoined her family. From her they learned that her companion still remained with the spirits in the peak—whether voluntarily or not is not clear.

One thing seems certain, that the earthborn woman was ready enough to come home when opportunity offered.

When the spirit women, who lived in the buttes and the trees, held their next annual gathering, they decided to found a chapter of their society in the Mandan village—perhaps out of friendship for their flesh-and-blood companion. She eagerly offered to go with them to act as their interpreter, since she now understood the language of spirits as well as that of Mandans. At this council in the peak, each of the spirits wore a bright green snake, such as are found in the buttes, for a headdress. The council must have resembled a circle of Medusas.

Having planned the new ritual which they were to confer upon their earthly sisters, they prepared for the trip. Being spirits, they were in the habit of walking around without touching the ground, and knew that they could have no difficulty in crossing the river. As for the Mandan woman, no doubt she was spirited across.

"When all was ready, the spirit women came out of the bare peak, crossed the Missouri River, and started for the Mandan village, still wearing the bright green snakes twined around their heads. After they had gone a little distance they met an eagle, who said, 'Let me go with you, and I will give you one of my feathers to add to your headdress.' The spirit women took one of the feathers and let the eagle come with them. For this reason a feather was always worn in the front of the headdress of this society. Next they met a coyote, who said, 'If I do not belong to your society it will not succeed.' The coyote gave them a song which was always sung at the close of the ceremony, and asked them to add to their headdress a wreath of plumy grass which resembled his fur. In return for this he was allowed to go with them and share the feast. As they came to a creek they met a bear, who said, 'You may meet trouble on your way, so you must wear claws to protect you from enemies you meet and from those who may follow you." The bear also gave them a song. For this reason the bear was allowed to go with them and share the feast, and when the ceremony was held there were two women who wore necklaces of bear's claws.

"The spirit women came to a creek and there they saw an otter and a flat clamshell. Both wanted to join the new society, so the spirit women allowed them to come. . . .

"When the spirit women entered the Mandan village they were still wearing the bright green snakes twined around their heads, together with the eagle feather and the wreath of plumy grass. They also wore the bear claws, the otter fur, and the polished shell. . . . They said, 'We bring the society because you are the people of the Missouri River,' and they told the young women to bring food for a feast. The young women took food and gifts into their lodge." [1]

Appropriately, the spirits of the Big Muddy were hearty creatures with good appetites. They steadily ate the food the Mandans put before them, but they did not talk out loud. They just sat there, bonneted with writhing snakes, uncannily whispering in each other's ears. That must have made the neophytes' skin crawl. One wonders whether the spirits sat on the ground or floated slightly above it.

After feasting they taught the Mandan women the songs of their society and "because they were spirit songs they were easy to learn. The maidens learned them all that night." They also taught their ceremony, and since the Mandan girls were naturally reluctant to wear live snakes in their hair, the spirits permitted them to braid grass to resemble snakes and so maintain their membership.

When the spirits had departed, the Mandan women set about their ceremony. It lasted four days, during which time the members remained together most of the day and slept in the ceremonial lodge at night. In the early evening the crier summoned them, and they paraded around the village to music which the men provided, afterward going to the lodge to dance and sing the songs taught them by the spirits and the animals who had been permitted to join. Apparently the music began slowly but became quite lively as time went on. Most of these songs were wordless. One, apparently referring to the original human member of the society, went something like this:

> Chum, where did you come from?
> I have come from a bare peak.

But the theme song of the society may be described as the Siren Song, and it may enable us to conjecture the general character or theme of the song Odysseus heard. It minced no words, but came right out with the central idea:

> The Missouri River women
> Are the best.

This song is believed to be the earliest slogan advertising products connected with the Missouri River. Perhaps publicity men along that stream, concentrating on real estate and industry, have overlooked something which the spirit women might teach them!

This sorority still exists and no doubt serves its purpose well. It is known, by some strange perversity, as the Little River Women Society. Membership in it is so valuable that young girls approaching marriageable age buy their way into it, paying a heavy initiation fee which has to include a buffalo robe or blanket and a good horse.

The claim to superiority of the Missouri River women does not, of course, rest wholly upon the virtues and self-confidence of the ladies of the Mandan village. If they choose, they can point to one woman who lived near them on that stream, a woman whose character and deeds have made her celebrated wherever the history of America is known. She has been honored by many public tributes, by the erection of several monuments and statues, and about her poems, songs, books, and innumerable articles have been written. She has been made the subject of not a few paintings.

This woman was Sacajawea, wife of Toussaint Charbonneau, who acted as guide and interpreter to Lewis and Clark from Fort Mandan on the Missouri to the Pacific coast and back, and so earned the undying gratitude of the people of the United States.

The Lewis and Clark expedition has been described as the most important single expedition ever made under the auspices of our government, and its story is "our national epic of exploration." It is perhaps not too much to say that without her help the expedition might have proved a failure.

Sacajawea was a daughter of the Snake, or Shoshone, Indians who lived to the west in the Rocky Mountains. But in 1800 the Hidatsa (Minnetarees or Gros Ventres of the Village) on one of their innumerable raids attacked the Snake camp at the Three Forks of the Missouri, killed four men, four women, and some boys, captured the remaining women and children, and carried them back to their palisaded town on the Big Muddy. Sacajawea was then only a girl. Apparently she first was the slave of the gigantic one-eyed chief, Le Borgne, described variously by travelers as "a Cyclops," a "monster," a "demon," a "cruel and villainous tyrant." He was probably the greatest scoundrel that ever lived on the banks of the Missouri. But she was soon purchased of Polyphemus by the Frenchman, Charbonneau. He brought her up, and when she became a woman, took her to wife, Indian-fashion.

During the winter (1804-05) spent at Fort Mandan, Lewis and Clark were much concerned to find guides who knew the country to the west. Their interpreter at the Mandan village, Jessaume, was not very competent and knew nothing of the language of the Indians on the headwaters of the Missouri. The two captains knew they could not cross the mountains in canoes or pack their heavy baggage over on their own backs. It was imperative that the party get horses from the Snakes, and to do that they *must* have an interpreter who well knew those Indians and could speak their language.

Fortunately Charbonneau learned of their difficulty, and arrived early in November with his young Indian wife to offer his services. He presented the explorers with four buffalo robes and made his request.

There was little reason to hire Charbonneau himself. But when Lewis and Clark learned that his wife was a Shoshone, or Snake, they lost no time in employing them both.

Sacajawea was a small, active woman, said to have been hardly more than five feet tall. At that time she was perhaps not twenty years of age, and was about to become a mother. As soon as this became apparent, Captain Lewis insisted that her feckless "husband" immediately marry the little captive in due form. The wedding took place on February 8, 1805, just in time to make the child legitimate. Already both the explorers had taken a liking to the little squaw—or "squar," as Clark wrote it. In those days Americans not only had freedom of speech but freedom of spelling.

On the 11th, when Sacajawea was in labor, the two captains behaved like distraught husbands, hovering about the house in which she lay and anxiously awaiting news of her condition. Captain Clark, after consulting Lewis, brought her a big belt made of blue glass beads, showed it to her, and hung it on a post beside her bed. Captain Lewis, equally concerned, on being told by Jessaume that a little dose of rattlesnake rattles would hasten the delivery, crumbled two rattles into a glass of water and gave it to her. Ten minutes later Sacajawea was the proud mother of a fine boy, Baptiste, better known as Pomp, the common Shoshone name for a first-born son.

By the time Pomp was three months old, the expedition was already well on its way up the Missouri River. Sacajawea, carrying her baby in the basket, or net, on her back, soon gained the respect and admiration of every man in the party, and lost no opportunity to show her appreciation of the consideration shown her by Lewis and Clark. And well she might. They were taking her home again, back to her own country and tribe. They had given better treatment to the poor captive squaw than any Indian woman dared hope for in those days. Once when the captains caught her husband abusing her they gave him a severe reprimand and warned him not to let it happen again.

Most of those who have left any record of Charbonneau seem to have thought him a foolish lout, born to blunder, always doing the wrong thing.

Quite early in the season Sacajawea had an opportunity to show her loyalty to her employers. On May 14th there was an accident. Needless to say, when the accident occurred, Charbonneau was steering the canoe. Clark records the event as follows:

". . . we proceeded on very well until about six o'clock. A squall of wind struck our sail broadside and turned the perogue nearly over, and in this situation the perogue remained until the sail was cut down, in which time she nearly filled with water. The articles which floated out were nearly all caught by the squaw who was in the rear. This accident had like to have cost us dearly; for in this perogue were embarked our papers, instruments, books, medicine, a great proportion of our merchandise, and, in short, almost every article indispensibly necessary to further the views and insure the success of the enterprise in which we are now launched to the distance of 2,200 miles."

Captain Lewis gives further details and pays his tribute to Sacajawea:

". . . By four o'clock in the evening our instruments, medicine, merchandise, provisions, were perfectly dried, repacked, and put on board the perogue. The loss we sustained was not so great as we had at first apprehended; our medicine sustained the greatest injury, several articles of which were entirely spoiled, and many considerably injured. The balance of our losses consisted of some garden seeds, a small quantity of gunpowder, and a few culinary articles which fell overboard and sunk. The Indian woman, to whom I ascribe equal fortitude and resolution with any person on board at the time of the accident caught and preserved most of the light articles which were washed overboard."

In recognition of Sacajawea's services the explorers a few days later named a river in her honor.

"About five miles above the mouth of Shell River, a handsome river about fifty yards in width discharged itself into the river on the starboard or upper side. This stream we called Sah-ca-ger-we-ah or 'Bird Woman's river,' after out interpreter, the Snake woman."

By the time the party reached the Great Falls of the Missouri, Sacajawea became "very sick." Captain Clark bled her, and then moved her into the back, covered part of the canoe to keep her out of the blazing sun. But her malady continued until she was in delirium. They gave her such medicine as they had, but made the—perhaps unavoidable—mistake of leaving her to the care of her husband. They were much worried—not only for her and for the child, but because she was their "only dependent for a friendly negotiation with the Snake Indians on whom we depend for horses to assist us in our portage from the Missouri to the Columbia River."

Fortunately they had just discovered the Great Sulphur Spring near the Falls. They made her drink the mineral water, which soon broke her fever and permitted her to take nourishment. Clark then gave Charbonneau specific orders as to her diet.

But as usual, Charbonneau was unequal to the trust reposed in him. A few days later she was able to get about and dug a quantity of "white apples" (*pommes blanches*) which she ate raw along with some dried fish. Her fever returned. Clark rebuked her husband severely.

A few days later Clark, his black man York, Charbonneau, and Sacajawea with her baby were out together when a storm cloud rushed upon them. A little way above the falls, Clark found a deep ravine where they could take shelter under some shelving rocks. They stowed most of their equipment under a ledge where the rain could not reach it. After a light shower, the cloud suddenly burst. Rain fell like a waterfall, instantly filling the ravine with a rushing torrent which carried rocks, mud, and everything before it. The flood threatened to sweep them into the river just above the great cataract, in which they must have perished.

Luckily Clark saw the torrent coming, snatched up his gun and shot pouch with his left hand and started up the bank. Sacajawea snatched up her child; Charbonneau had already gone ahead to make himself safe. Clark pushed the Indian woman before him, and no doubt called to her husband to help. Charbonneau turned and took her hand, but was too scared to be of any use. Clark, wet to the waist in the sudden rise, managed to save the woman and her child.

Sacajawea had lost her baby basket and all the child's clothing. The wind was cold and she, a convalescent, stood drenched and shivering. Below them they saw the furious torrent, now fifteen feet deep, which had swept away Clark's compass and his umbrella.

Fortunately York, the Negro, had a canteen with some spirits in it. Clark made Sacajawea drink. They then all ran back to camp where they could find dry clothing and shelter. . . .

All along the way Sacajawea continued to serve her employers. She knew how to cook marrowbones; how to find the hoards of wild artichokes buried by rodents; and sometimes she gathered "a fine parcel of service berries."

As they went on, she recognized the country, pointing

out the Beaverhead and other landmarks. At Three Forks she showed them the place where she had been captured as a girl. Lewis was surprised to find that she showed no emotion in describing that event and no joy in being restored to her native country. Says he, "If she has enough to eat and a few trinkets to wear, I believe she will be perfectly content anywhere."

But the captain was mistaken in thinking this young matron without emotion. For as she drew near the home of her people in the company of her husband and Captain Clark, she "suddenly began to dance and show every mark of the most extravagant joy, pointing to a group of mounted Indians approaching. She sucked her fingers to indicate that they were her native tribe." This gesture (from the sign language of the Plains Indians) of putting two or more fingers into the mouth is intended to indicate that the persons so referred to had been suckled by the same mother or were blood relatives.

The Snakes sat in council with Clark and agreed to sell horses to the explorers. Sacajawea interpreted to her husband, who translated again for the captain. Suddenly Sacajawea jumped up and hugged the chief, throwing her blanket about him and bursting into tears. She had recognized her long-lost brother.

The chief on his part also showed signs of affection, but evidently thought that a council was no place for such demonstrations. He made his sister sit down again. Still she had difficulty in interpreting for him. Whenever he spoke, she wept. All her family were dead now except for two brothers and the son of her eldest sister. This boy she immediately adopted.

Nevertheless, Sacajawea's love for her brother did not lessen her loyalty to the explorers. She informed them that the chief, after agreeing to supply them with horses, had ordered his people from the mountains to the plains to hunt buffalo. Probably he wished to kill some meat before he parted with his animals, not realizing that time was of the essence of his contract to supply the explorers.

Clark was flabbergasted. It seemed the whole expedition must fall through. He assembled the chiefs, reminded them of their promise to sell him horses, and asked if they were men of their word. The chief made apologies and countermanded his order. Sacajawea had saved the day. This single incident would be quite enough to justify the admiration and gratitude felt by all true Americans for "our Indian woman."

At the Snake village a new difficulty beset her. It was the custom among the Snakes for a father to promise his daughter before she became old enough to marry, and Sacajawea, as a child, had been so promised to a Snake warrior. The fellow was more than double her age and had two other wives; but, having paid for her, he came forward to claim her. However, on seeing her child by Charbonneau, he withdrew his claim. Sacajawea went on with the party. It is pleasing to recall the appreciation shown by the explorers for her services at this time. They gave her husband trade goods enough to buy a horse for her, so that she could ride.

A story popular all over the plains tells how an Indian buck was riding comfortably along on his pony while his squaw followed on foot, carrying the baggage. A white man stopped the Indian and demanded, "Hey, John, why don't the woman ride?"

Soberly the buck replied, "She got no horse."

From the Indian point of view, this was no joke, for a horse was individual property. There was no "family horse." If the woman had owned the horse, you can bet her husband would have had to walk.

On the Missouri, Indian women were often independent as a hog on ice and sticklers for their rights. Bull Head's comical misadventures with his spouse will illustrate the spirit of some of these prairie women:

"We were going toward the lower Missouri when the men went hunting. Bull Head was along with them. He was coming in with his meat when, looking down from the top of a hill, he saw his wife riding along with the other women, sitting on top of her pack and dragging a travois with their camp goods. He rode down and gave her the meat to pack into camp. While she was transferring the meat from his horse to hers, he sat down to wait. Just when she crossed behind the pack horse, it kicked out and hit her hard. She went down, and the horse stampeded. Her child burst out wailing, 'My mother is killed.'

"Bull Head came running, all solicitous. 'Where are you hurt? Where did he kick you?'

"'Keep away from me,' his wife screamed. 'Don't you come near me. You men! You lazy fellows! It is not far to the camp, and yet you make me pack the meat. It is all your fault.'

"But Bull Head did not heed her as he went to help her up. She just took his war spear and broke it over his shoulders, wherewith he went off to look after the runaway horse, while she came hobbling after.

"They found the horse mired in a mudhole. 'There, see what you did,' she yelled. 'You stay away.' Then she tried to get the lead rope, but could not reach it. She ordered her repentent husband into the mudhole to get it himself. He obeyed her, but the horse reared, stepping on his foot. He disappeared from sight under the water, going right under the horse's belly and coming up on the other side—a mess. His wife plucked a handful of mud and hurled it just as Bull Head turned his face. It smeared him; he threw some back—and they were at it.

"In the meantime the horse was drowning. Hawk rode up, and while they were fighting, he lassoed the horse and pulled it out. Then he yelled to them to help him and stop ducking each other. When it was over, the meat was ruined.

"When they came into the camp late, Sharp Nose, an old woman, invited Bull Head to stop and eat. 'Why don't you wash your face?' she asked him curiously.

"'I did,' was his reply. But no, he had only smeared the mud around.

"Then the daughter came over there.

"'What is the matter with your parents?' was Sharp Nose's query. 'Everyone has her lodge up, but your mother has nothing.'

"'The truth is, that wife of his refused to put up the tipi or cook for him.' Her mother tried to soften her, saying, 'Maybe you are wrong.'

"But then she remained stubborn, 'Not until my arm is well will I do a thing for him.'

"Whenever the mother-in-law gave the little girl a bit of food to take to her father, the injured wife seized it and threw it to the dogs.

"Bull Head had to go rustling food from camp to camp. 'I am being punished,' he told the people. 'My wife won't feed me.'

"But after some time his wife took him back. This man had a good war reputation and a great medicine for curing snake bites. . . ." [2]

Sacajawea went on with the explorers clear to the Pacific Ocean. There, in order to obtain a fine robe of otterskins for one of the captains, she parted with her treasured belt of blue beads. A true Indian, the little squaw took delight in traveling.

One day the carcass of a whale was discovered on the beach. Clark and others went down to see it. She begged that she might go along, pleading that she had traveled a great way to see the Great Water yet had never been down to the beach, and now that this monstrous fish was also to be seen, it seemed hard that she could be permitted to see neither the ocean nor the whale. Clark took her along. Of course, it had never occurred to her husband to do such a thing.

Returning from the coast, the party divided. Clark had taken a great liking to Sacajawea's son Pomp. On his way down the Yellowstone he honored the boy by giving his name to the principal landmark on that stream—Pomp's Tower, now absurdly called Pompey's Pillar. Inasmuch as this butte does not resemble a pillar in any respect and certainly has nothing whatever to do with Pompey, it is a shame that the original name has not been restored to the map. Sacajawea's name also fared badly at the hands of later mapmakers. The river named for her is now called Crooked Creek!

Apparently Clark took great delight in watching her little boy "dance," and his interest in the child in later years showed how fond he had become of the small warrior. Indeed, the presence of this admirable woman and her child on the expedition must have meant a great deal to all those lonely men so far from home.

Back at the Hidatsa village on the Missouri River, the explorers prepared to start home. Their journals devote a paragraph to the parting. From this it appears that Clark offered to take Pomp, "a beautiful promising child," to the settlements and raise him in a civilized manner, and that the parents agreed that he might do so as soon as the child was weaned. And it appears from the Biddle edition (August 16th) that Clark would gladly have taken the whole family home with him:

"The principal chiefs of the Minnetarees came down to bid us farewell, as none of them could be prevailed on to go with us. This circumstance induced our interpreter, Chaboneau, with his wife and child, to remain here, as he could no longer be useful; and notwithstanding our offers of taking him with us to the United States, he said that he had there no acquaintance and no chance of making a livelihood, and preferred remaining among the Indians. This man has been very serviceable to us, and his wife particularly useful among the Shoshonees. Indeed, she has borne with a patience truly admirable the fatigues of so long a route incumbered with the charge of an infant, who is even now only nineteen months old. We therefore paid him his wages, amounting to five hundred dollars and thirty-three cents, including the price of a horse and a lodge (a leather tent) purchased of him."

In a letter written four days later to Charbonneau, Clark renewed his offer, promised to find employment or a farm for the Frenchman if he would only leave "your little son (my boy Pomp)" so that Clark might "educate him and treat him as my own child."

From the same letter it appears that Sacajawea was known familiarly to the explorers as "Janey!" The name Sacajawea has been interpreted in various ways. Some think it an Hidatsa name, Tsakáka-Wia, meaning Bird Woman. Others think it a Snake name meaning Boat Woman or Boat Launcher.

Afterward, as Indians will, she had other names. For Clark was as good as his word. Sacajawea and her son went down to St. Louis, where the boy was put to school. Afterward the "little dancing boy" Pomp, or Baptiste, became a member of the household of Prince Paul of Württemberg, and accompanied him to Europe in 1825.

From St. Louis, the story goes, his mother later went to what is now Oklahoma, married a Comanche, and spent a number of years with that tribe, one cognate to her own people. Even in her old age they say she was a handsome little woman and one whom everybody noticed. In fact, her biographer, Grace Raymond Hebard records that, when she first came among them, the envious Comanche women dubbed her "Flirt!" ³

But eventually, it is said, Janey returned to her own people and lived to an honored and ripe old age on their reservation in Wyoming. There, they say, she lived alone in her tipi near Fort Washakie. And there, in her sleep, on the night preceding April 9, 1884, she died.

CHAPTER XXII

The Little Missouri

A HUNDRED years ago the name Little Missouri was given to the Bad, or Teton, River in South Dakota, but today it refers to a small, winding, muddy stream which heads near the Black Hills and flows north along the western edge of North Dakota, turning east to strike the Missouri a few miles above Elbowoods. No stranger or more lonely valley can be found on the Plains. In fact, there is nothing quite like it in North America.

The Little Missouri suggests some celebrated names.

Of these, that of Wild Bill Hickok, gambler, gunman, and peace officer in Deadwood, may best represent the Black Hills. His grave lies high above the town alongside that of Calamity Jane. One wonders how they ever got the coffins up there.

DEADWOOD

"Oh, who is that beside the bar,
So handsome and so tall,
With gray sombrero, high-heeled boots,
And the black cape over all?
With his taper lady fingers,
With his long silken hair,
His long-tailed coat and flower-bed vest? . . ."
    "Wild Bill is standing there!

"He has killed a hundred gun-men
On duty or in self-defense,

WILD BILL HICKOK

And he buried every man of them
At his own expense.
He started many a graveyard,
He made men fear the Law—
Sure of eye, and cool of head,
And lightning on the draw!"

"Oh, who is that outside the door,
That shabby, cross-eyed runt?
Oh, who is that comes sneaking in
Like a weasel on the hunt?
While Bill is playing poker
He slips behind Bill's chair—
McCall, the tinhorn gambler—
    What is he doing there?"

Wild Bill has never a warning,
McCall behind him stands—
Yet as the ball plows through his brain,
Bill's guns are in his hands!
Quick as a wink he draws to shoot—
Not Death itself was half so fast—
Wild Bill lies dead in his high-heeled boots,
Unbeaten to the last!

So men will long remember
Wild Bill in Deadwood town,

And many a lad will curse the coward
Who shot his hero down;
For always now, in Deadwood,
Among the bold Black Hills,
They show the stranger a famous grave—
Brave Wild Bill's.

After the Little Missouri enters South Dakota, the first town of importance is Marmarth, with about seven hundred inhabitants. This "city of trees," often flooded by the stream, may seem only a village to the stranger; but in that country a town is important, not for its human population, but for the number of cattle it ships to market. The stockyards at Marmarth cover forty-five acres.

Nearby are the crumbling ruins of old Fort Dilts, where a handful of pioneers defended themselves against the Sioux behind ramparts of sod cut from the prairie. The river winds northward through the Badlands to Medora, the seat of the Marquis de Mores. There one may still see his "château" and his bronze statue, together with the fire-marked ruins of the packing plant in which he sank his fortune in the eighties. De Mores was the nobleman whom Theodore Roosevelt, then a young asthmatic hardening himself in the wilds, once challenged to a duel.

The Marquis de Mores prided himself upon his distinguished ancestry. He honestly loved the Little Missouri country and took great pride in the town and industry he had founded at Medora. But his alien background and his tradition of special privilege made it hard for him to understand or sympathize with the rough democracy of the frontier. He thought of himself as a feudal lord, and though very generous to his retainers and willing when necessary to risk his skin in their defense, he was a poor judge of men and sought advice in the wrong quarters. When a newspaper was started in his town, he was astonished to find that the editor insisted upon absolute freedom of the press and proposed to finance the paper himself.

When some of his herdsmen ran afoul of three professional hunters, the marquis appealed to a local magistrate, asking what he should do. The judge advised him to shoot. Not understanding the implications of that advice in that country, the marquis and his men ambushed the hunters. When the shooting was over, one of the three lay dead. The two surviving hunters were arrested and charged with manslaughter.

Later, the marquis himself was indicted for murder in the first degree; though acquitted, he found afterward he was very unpopular in the district. Mysteriously all his projects were sabotaged, and as time went on he came to blame Theodore Roosevelt for his troubles.

Now Roosevelt, like the marquis, was an aristocrat and had high ambitions, but he was in thorough sympathy with his neighbors and their democratic ideals. Accordingly, when the French noble wrote Roosevelt a nasty letter and suggested that there was "a way of settling differences between gentlemen," Teddy immediately declared, "He can't bully me!"

Though heartily disapproving of duels, Teddy immediately sent an answer stating that, since the marquis had threatened him, he was ready to answer for his actions. With characteristic directness, he did not wait for an acceptance of the challenge, but informed the Frenchman that he chose "rifles at twelve paces, the adversaries to shoot and advance until one or the other dropped." [1]

But the marquis was no coward, and, not having intended to bring on a duel, was honest enough to admit it. He requested Roosevelt to do him the honor to dine with him. That ended the difficulty.

Teddy Roosevelt, on his first hunting trip to the Badlands, fell in love with the region, and went into the cattle

THEODORE
ROOSEVELT

business there. Though he lost money, he had a wonderful time in that hunter's paradise, which in those days swarmed with otter and beaver, deer, elk, antelope, buffalo, grizzlies, mountain lions, and bighorn sheep—to say nothing of bobcats, coyotes, and prairie dogs. On the Little Missouri, Teddy also became an impromptu peace officer, rounded up three bandits and brought them to justice. It is not too much to say that the Little Missouri made a man of Roosevelt. Certainly his years there gave him that understanding of frontier life which endeared him to all Westerners and made him the hero of the Rough Riders in the Spanish-American War.

Even today that river is wild and unsettled, for the stream flows through country so broken, tumbled, and confused that only a cowboy or an Indian could find his way around and survive there.

The Badlands of the Little Missouri cover hundreds of square miles with one of the most fantastic landscapes in North America, a welter of strange ridges and bare hillocks, pyramids, domes, and buttes, barred with horizontal stripes of varicolored strata, red, yellow, brown, and gray, which contrast beautifully with the silver sage below and the dark cedars on the eroded slopes. They form a labyrinth of waterless dead hills and dry coulees—an unearthly country, fantastic and tumultuous—at once serene and incredible.

Yet among these strange hills, in spring and summer, the valleys show a riot of wild flowers; primroses, lilies, chokecherries, flowering currants, and prickly pear.

Along the crooked stream one finds groves of cottonwood, elm, ash, box elder, pine, and dogwood, so that the Sioux knew the stream as Thick Timber River. It was one of their favorite winter camp grounds where, in the midst of buffalo and other game, they could pitch their snug, smoke-browned tipis under the shelter of the painted bluffs.

General Alfred Sully, on his arduous campaign against the tribe in 1864, described that weird country as "Hell with the fires out."

But this description of Sully's is not so accurate as its popularity would indicate. The fires are not out yet. For all through that country one sees buttes scorched pink or red by the heat of burning coal veins. Even today some veins of lignite still burn underground. In summer they are not easily found, but in cold weather the heat makes steam rise into the air to show that the Devil is still at work in his subterranean mines.

There are thousands of petrified trees in the Badlands, and innumerable fossil skeletons of prehistoric animals—saber-toothed tigers and giant hogs. The Grand Canyon of the Little Missouri is something to behold.

Almost every traveler since Sully has burst into ecstatic praise of the strange charm and beauty of the region. It is painted in dazzling white, flaming scarlet, brown, buff, lavender, and gray, besides a thousand pastel shades. Some of the buttes exactly resemble in color and form a layer cake, a tasty muffin, or a loaf of home-baked bread. It was tough country for soldiers and cavalry horses, but for the man with an eye for beauty and a taste for solitude it seems a genuine garden of God.

A whole book could be filled with stories of Indian hunts and battles on the Little Missouri, some with other redskins, some with soldiers of the United States. And these fights were sometimes as unusual as the bizarre country in which they took place: The Battle of the Badlands, The Battle of Killdeer Mountain.

My old friend, Chief White Bull, nephew of Sitting Bull, took part as a boy in the Battle of Killdeer Mountain. I have cast one of his stories of the fight into the ballad that follows:

BEAR'S HEART

General Sully at Killdeer Mountain
He shelled the tipis of the Sioux;
Their thin red line fell slowly back
Before the coats of blue.

Shells were bursting among the tipis,
Women and children were on the run,
It seemed Sioux arrows could not defend them
'Gainst long-range rifle, and long-range gun . . .

Up from the rear a man came riding,
Leading an old gray horse with a drag;
A huddled figure lay in the basket
Behind the old gray nag.

They called him The-Man-Who-Never-Walked,
No honor-feathers were in his hair;
From birth his limbs had been wried and withered,
But he had the heart of a bear.

He looked at the smoke of the coming soldiers,
He heard the crash of the bursting shell;
The flash and rattle of barking carbines,
The whine of bullets pleased him well.

"I have no strength to draw the bow,
I never could fight as the others can;

I cry in my heart that I lived like a woman,
Let me die the death of a man!"

Sitting Bull's heart was big with pity,
His heart was big with pride in his own—
"Let him have his wish to die in battle,
Let him make his charge—alone."

He laid his lash on the gray nag's withers,
The gray sped off with the man behind;
Straight for the smoke of the flashing rifles,
Straight as an arrow down the wind.

Sudden the horse drops dead as a boulder,
Out of the basket the cripple falls;
With a whoop and a song he sat on the prairie—
He could not dodge the balls.

The rifles' roar soon drowned his music,
The rifle bullets soon stopped his cry;
But Sitting Bull's Sioux fell back together—
Bear's Heart had showed them how to die.

CHAPTER XXIII

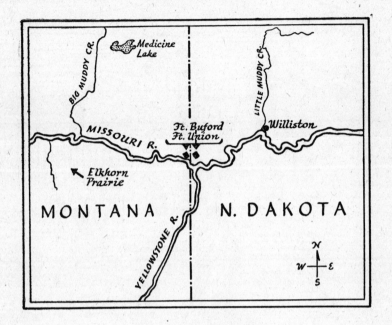

Fort Union

SETTLERS on the Missouri River, like the people of Boston, built their homes along the cowpaths. The severe winters on the Upper Missouri drove the buffalo cows to take shelter in the valley, and the Indians swarmed in after the cows. Where the Indians were to be found was just the place for traders, Indian agents, and missionaries; and so the military posts, later established to protect the agencies, trading posts, and missions, were inevitably built upon the river.

FORT UNION, LARGEST TRADING POST ON THE

RIVER, STOOD AT THE MOUTH OF THE YELLOWSTONE.

Boatmen naturally made landings at such points, and when the railroads came in, these touched the river where settlements already existed. The cities which arose where the iron rails leaped the river are there now because buffalo cows broke the trail.

Thus the Upper River has always been more populous than the vast plains through which it flows. An incredible number of forts, posts, and "houses" devoted to the fur trade were established on the river. In fact, they were more numerous a century ago than the towns which have succeeded them are today. Most of these "forts" were small affairs consisting of a few log houses, quickly built and soon abandoned. Some were great establishments occupied by a considerable population for many years.

Such was Fort Pierre, named after Pierre Chouteau, at the mouth of Teton River opposite the present capital of South Dakota; this post traded with the Sioux. Farther up the river at the Arikara village stood Fort Clark. The Hidatsa, or Gros Ventres of the Village were served by Fort Berthold, a few miles below the mouth of the Little Missouri. Fort McKenzie stood near the mouth of Marias River; it traded with the Blackfeet. All these were important in the life and history of the stream.

But the largest trading post on the river stood at the mouth of the Yellowstone. Fort Union was built in 1829 by Kenneth McKenzie. Coming up the river on the annual steamboat, men could see the fort, with its flag flying and its stout bastions, standing miles away on the wide prairie bench. As the boat approached the landing, cannon from the fort would thunder, while those afloat replied by firing all their guns.

Once ashore, the traveler approached the spacious stockade of square hand-hewn pickets twenty feet high, stretching for a distance of some 240 feet. Entering through the stout double gates, he found himself in a courtyard almost as wide as it was long, with bastions twenty-four feet square and two stories high, built entirely of stone and having a pyramidal roof, at opposite corners, for defense. Across from the gate stood the handsome and commodious two-story house of the bourgeois, with two stone chimneys, glass windows, shutters, and a gallery or piazza in front adorned with turned pillars painted white. The other sides of the fort were occupied by the barracks of the engagés or employees, the warehouses, workshops, stables, fur press, and stone powder magazine. In the midst of the court a handsome flagstaff rose high above the walls—a tall stem on which Old Glory flowered. The buildings were all of cottonwood timber, and everything was in apple-pie order.

When admitted to McKenzie's house, the guest opened his eyes. For here, at this remote crossroads of the wilderness, he saw walls covered with real wallpaper, rooms with carpets and fine furniture, a dining table groaning under buffalo hump ribs and pickled tongues, beaver tails and marrow fat, fresh vegetables from the post garden, venison and elk steaks, with a bucket of ice alongside holding one bottle of choice Madeira and another of excellent port. Quite possibly McKenzie came to greet him dressed in the fine red coat which he wore to impress the Indians. McKenzie lived like a lord, sharing his home with an English gentleman named Hamilton. That bourgeois well deserved his title, "King of the Missouri."

The visitor, finding his way about the establishment, soon discovered the store at the gate—a long room with a counter running the length of it down the middle, so that none of the customers could reach the goods on the shelves behind it—and inspected the tailorshop, saddlery, icehouse, and storerooms, and made the acquaintance of forty or fifty white men, half-breeds, and full-bloods, employed there. He saw the artillery in the bastions and noted that the brass cannon beside the flagstaff stood trained on the gate.

Life at Fort Union between boats might seem uneventful and even monotonous. For the newspapers, carefully hoarded and read in chronological order, gave only the news of the year before, while letters came only on the annual steamboat, or overland by special express from St. Louis.

Fort Union was exposed to all the winds that blew, and was none too warm in winter. Still, it was not without comfort, amusement, and even excitement, when Indian warriors covered from top to toe with the black paint of victory came in to show their scalps and dance in triumph, when the prairie outside was covered with the smoke-browned tipis of the Assiniboins, or when a party of trappers came in to dispose of their furs and spend their earnings in a wild debauch. A nosy visitor might even smell out the secret distillery on the premises which made the forbidden liquor on which the fort's Indian trade depended.

Attached to the building, but outside it, was a reception room for visiting Indians, who could there find shelter without actually getting inside the stockade. And some miles up the stream was the chantier, or "navy yard," where boats were built and fuel cut for the fires.

The people at the post had frequent dances at which Indian women assisted. Culbertson could play the fiddle, Mr. Guépe the clarinet, and Mr. Chouteau the drum. Cotillions and reels were performed with great energy and enjoyment, and the dances broke up in the wee small hours. There were always riding parties, picnics, and hunts to be enjoyed. Men like McKenzie and Culbertson were crack shots and excellent horsemen. Audubon tells how Culbertson, on a bet, mounted his horse, pursued a wolf on the prairie, killed it, lifted it across his saddle, and delivered the carcass to the naturalist within the space of twenty minutes. Numerous stories are told of McKenzie's prowess in the buffalo hunt.

The profits of the fur trade were enormous, for both parties gave up what cost them little for valuable goods which neither could manufacture for himself. All that is an old story.

There was one aspect of the fur trade, however, which has generally been overlooked. Some have complained that the history of the Missouri River is lacking in bedroom scenes, and that the stream affords few examples of what are popularly known as "great lovers." Of course, the great lovers on any stream are really those men who care enough for their mates to stick to them and faithfully raise their families in their homes, and millions of these have lived and died on the stream. One of the head men at Fort Union is an example. He had a fullblood Blackfoot wife, a woman of great intelligence and high character, who on occasion wore European costume and made an admirable wife—though visitors deplored her fondness for raw buffalo gut.

But in those early days, when polygamy was the custom among the Indians and women far outnumbered men, the amorous exploits of some of the frontiersmen would make Casanova and Don Juan look like amateurs. In the records of the fur trade there are numerous references to the affairs of such men. For there was something about the fur trade which encouraged trickery of every sort, and where all men carried arms, few were much given to interfering with other men's enterprises.

The artist, Kurz, has recorded in his journal a typical exploit:

"Sitting before a jolly open fire, by the side of the wounded Packinaud, while a cold, strong wind howled in the courtyard without, I listened to amusing adventures related by Carafel, who was today in fine humor. One of his stories interested me more than the others because I know most of the people concerned. The hero is that sly dog Vilandre. . . . He is equally well-known for a tiptop trapper, trader, and hunter, and for a reckless spendthrift not overscrupulous as to honesty. One day old Gre and his comrade hired Vilandre to go along with them and trap beavers in . . . Blackfeet territory. On the way his two masters quarreled all the time, never agreed with each other about anything, and finally brought matters to such a pass that our hero could stand it no longer. He said to them:

" 'Now, here, this won't do. You two never agree; after this fashion we shall accomplish nothing. Two masters for one hired man is a bad arrangement; it is better for me to employ both of you. Sell me your horses and traps, and I will pay you later on in beaver skins.'

"They traded. From the hunt they realized an unusually large profit; it was, you see, in the good old days. They brought back between 400 and 500 bundles of ten skins each.

"Colonel Mitchell, the present U. S. agent of the Indian Department in St. Louis, was then *bourgeois* at Fort Union. When he saw the three trappers approaching, he went forward to meet the two old traders, paying no attention to Vilandre, who was leading the way with two heavily laden horses. He knew nothing of the later agreement and intended, of course, to make himself agreeable to those who went out in charge of affairs, congratulate them upon their successful hunt, and invite them in. Vilandre passed on by the employees of the company who were standing beside the gate, curiously looking on, while waiting to receive the packets. He carried his head as high as did the *bourgeois* himself. And imagine the astonishment of the latter when he saw Vilandre haughtily pass the gate!

" 'What! You will not trade with me?' he asked the old 'grumblers.'

" 'The beaver skins belong to Vilandre, not to us.'

"The expressions of certain faces immediately changed —business took a different turn altogether. Mr. Denig, clerk at that time, was dispatched at once to present his compliments to Vilandre and ask him to unload his pelts here. In the end he relented. As he rode up to the gate he caught sight of a splendid girl, daughter of old Garion by an Indian woman. He was much pleased with her, and said to himself. 'This night you shall be mine.'

"He sold his furs, which were bringing high prices at that time, paid his quasi-employers forthwith, clothed himself throughout in new apparel, assumed the arrogance of the devil, and bought a great number of articles for gifts. When the horses were brought in from the pasture in the evening and corraled, he took his best runner and leading him to Garion, said brusquely, 'Take him. He is yours!' Garion, suspecting at once what Vilandre's purpose was, demurred. 'Holy Virgin,' shouted Vilandre, 'am I such a poor man that I can't afford to give away a horse? Haven't I still my little gray?'

"Thereupon he stroked the gray horse caressingly, without bestowing even a glance upon the beautiful girl near him. But he sent a friend a little later to Garion asking his daughter in marriage. Her father consented on condition that a marriage contract be drawn up in writing and signed. Well and good. Vilandre went at once to Moncrevier (now with the Pawnee), requesting the draft of an agreement that, when read aloud, would contain all the usual expressions and requirements, but with this difference: A provision was to be inserted permitting him to leave the bride whenever he chose.

"Moncrevier, a sort of wag, found much pleasure in anticipation of the impending wedding feast and made all things ready. Garion was satisfied, his daughter, who received many presents, was content; and Vilandre on the same night, after his return to the fort, was in possession of his beautiful bride.

"How long do you think he kept her? One entire winter. When he determined to send her away he set her on a good horse and presented her with abundant apparel as well as provisions.

" 'Holy Virgin,' he cried, 'when I bid farewell to my wife, whom I am casting off, I will not take the shirt from her back. I put her on a horse and supply her with goods. I don't mind the loss of a beaker of coffee.' " [1]

Sometimes the trickery of the fur traders was a savage thing.

Of these bloody crimes, several occurred at Fort McKenzie, where in 1832 Prince Maximilian witnessed the battle when the Assiniboins attacked the Blackfeet camped about the fort.

Here, too, ten years later, one of the most villainous crimes in the history of the Missouri was committed by the man in charge, Chardon, and Alexander Harvey. Some Blood Indians had killed a Negro belonging to Chardon. The trader had been very fond of his black servant and selected Harvey, one of the toughest desperadoes on the river, to help him wreak vengeance on the Indians.

For Harvey was already notorious for sanguinary deeds. In that very fort he had quarreled with a Spanish employee only two years before. Harvey challenged the Spaniard to shoot it out, but the man would not take up the challenge. So Harvey, standing in the store, whipped out his gun, and saying, "You won't fight me like a man, so take that," shot him through the head. Then Harvey took his stand in the middle of the courtyard and called out, "I have killed the Spaniard. If any of his friends want to take it up, here I am." But no one had dared oppose him.

So now Chardon and Harvey planned vengeance. They loaded the cannon with rifle bullets and trained it on the gate of the fort, waiting for the Bloods to come in to trade. When they approached, Harvey took his stand by the cannon with a lighted match, while Chardon, armed with a rifle, lay in wait for the chiefs.

Whether Chardon had been drinking, or was too quick on the trigger, he opened fire before Harvey was ready.

Chardon dropped one chief at the first shot, but before Harvey could put fire to the touchhole, the Indians in the gate had scattered. Only five were hit; only two were killed.

The two murderers had expected rich booty, but the Indians ran off with all their furs and saddles.

Furious at his disappointment, Harvey finished off the wounded with his knife, scalped the dead, and in savage humor made the Indian women dance over the hair torn from their sons and husbands.

Of course, this made further trading at the fort impossible. Chardon pulled out. The Indians burned the fort. Its site is known to this day as Brulé Bottom.

CHAPTER XXIV

The Great Dam

THE Upper Missouri is not without its stories of hidden treasure. Somewhere about the mouth of Burnt Creek $20,000 in gold dust was lost in the summer of 1863. A party of white people, including at least one woman, a young girl, and a baby, were returning to the States from the Montana mines. Among them they had nearly $100,000 in gold dust packed in money belts and pouches and in holes drilled into the planks of their Mackinaw boat.

At Fort Berthold the trader Gerard warned them that the Sioux were hostile. General Sibley had chased the Indians across the Missouri, but they had come back and were hunting on the east bank not far below.

But the miners distrusted Gerard. They had heard what skinflints Indian traders were. They thought he wanted to keep them at the post where they would have to buy supplies from him at his own price. The next morning they piled into their boat and steered downriver.

At the mouth of Burnt Creek lay a long sandbar with a shallow, narrow channel under the bank. As the Mackinaw slipped down to the head of the bar, the white men saw an old Indian fishing on the bar. Whether out of friendliness or because he did not wish the fish to be disturbed, the old man waved them away from the narrow channel, signaling them to keep to the main current on the outside of the bar.

In spite of their disdain of Gerard's warning, the

miners now became alarmed. They thought the Indian was signaling to warriors on the bank; they fired and killed him.

A number of Indian women, then bathing near the mouth of the creek, rushed back to their camp, squalling and crying.

Down came the warriors to the river, but the white men were ready. They had brought along a small cannon loaded with rifle bullets. They touched it off and knocked over several Indians.

But at the second or third shot, their cannon kicked a hole in the boat, which promptly sank, lodging on the sandbar beneath the shallow water.

In this pretty fix it was not long before the white men began to get the worst of it. Their leader was shot. The Sioux rushed in and butchered the others to the last man.

In looting the boat they found the bags and belts of gold dust. Not knowing its value, they ripped up the sacks and scattered the dust around the bar.

Some days later a party of Indians brought the news to the fort. Gerard immediately persuaded his brother-in-law, Whistling Bear, to organize a party of Mandans to go and recover the gold.

The Mandans found the boat and in it an old coffee-pot. With this they scooped up all the gold they could find on the bar and returned to Fort Berthold. Gerard must have rubbed his hands—they brought back $70,000 in gold dust, and $70,000 was money in those days even for an Indian trader. To show his appreciation, Gerard gave his brother-in-law a fine horse—probably worth $100—and made a feast for the other members of the party—probably costing $10.

Later, when it was learned that gold had been hidden in holes drilled in the planks of the boat, another search was made; but in the meantime the river had risen and swept the boat away. In the past there have been numbers of men who poked around the mouth of Burnt Creek and below it in the hopes of finding those precious planks. If any of them ever found anything, no record of it has been published. In those days they wasted their time looking for gold in the Missouri.

But recently people along the Missouri have made a gold strike which completely dwarfs even the old pioneer gold rushes on its headwaters. They have struck pay dirt in the biggest gold mine in the world—the United States Treasury, from which the river drains an endless flow of wealth.

For more than a hundred years the national government has been spending money on Missouri River development, and vast sums of money have been expended in making soundings and mapping the main stem, in snagging, dredging channels, laying revetments, building cribs. Finally, under the New Deal, the waters of the Missouri were impounded by the Fort Peck Dam. The dam cost about sixty million dollars, and $60,000,000 looked as big as the state of Montana to a citizen of Fort Peck.

Dazzled by the scale of this project, eight states, Missouri, Kansas, Iowa, Nebraska, South Dakota, North Dakota, Montana, and Wyoming, have banded together in a Missouri River States Committee in Congress for the express purpose of securing a valley-wide development of the entire river system for the multiple purposes of power development, irrigation, flood control, navigation, and related improvements. Instead of each one depending upon piddling local projects, these states have formed a united front to co-operate in securing federal funds.

According to their spokesmen in Congress, they are seeking something similar to the Tennessee Valley, Colorado River, Columbia River, and other great river valley developments: "The general idea is that a series of multiple-purpose dams below Fort Peck plus dams on the tributaries and a series of levees in the lower reaches of the river would provide an effective plan of flood control and at the same time furnish aids for navigation, irrigation, and power development." [1]

Those who favor this undertaking maintain that the project would be self-sustaining and pay for itself.

So vast a program, of course, is not to be completed all at once. If the taxpayers permit, it will probably be continued (to use the formula of old Indian treaties) "so long as grass grows and water runs."

The problem of engineers on the Missouri is quite simple—if *not* easy: it is just to get the river to move water instead of earth. This is not likely to be accomplished in our time. It is a huge task.

But it is fitting that men on the Missouri should dream great dreams. The puny undertakings suited to small, tame streams do not belong on the Big Muddy. And indeed, the first dam built across the river at Fort Peck, Montana, is worthy of the stream. In fact, it is so big that you may overlook it. You go driving along in your car below it, looking for the dam. Then you happen to look up, and are appalled to realize that this enormous hill you are skirting is the dam itself.

That dam towers 242 feet above its base, and is nearly half a mile wide at the bottom. The bluffs on either side which it connects are almost two miles apart, and in addition a four-mile dike was built on the west side to keep water from running around the end of the dam.

The river had no bedrock at this natural bottleneck where the hills approach the river—no bedrock to which a masonry dam could be anchored or upon which a rock-fill dam could be based. Accordingly, the Fort Peck Dam is an *earth-fill dam*. To build it 130,000,000 cubic yards of dirt was moved, 4,000,000 cubic yards of gravel, and 1,600,000 cubic yards of riprap was laid down. The dam is an hydraulic fill made by pumping mud and water raised by four dredging boats through steel pipe lines, so that the gravel and sand and clay remained while the water, with its fine silt, ran back into the river. Thus were built two porous sand and gravel "shells," a core of clay and fine sand.

Four diversion channels, or tunnels, were dug through the hills to carry off surplus water. These tunnels, each 24 feet wide, burrow through the rocky bluffs east of the river for more than a mile and empty well below the dam.

The reservoir, or lake, behind the dam extends up the Missouri for 180 miles. It is 16 miles across at its widest, and holds enough water to cover the entire state of Montana $2\frac{1}{2}$ inches deep. Montana is a huge state, and that part of it is fully as "roomy" as any other; yet, even for Montana, the Fort Peck Dam is BIG.

And it was a big job to build it. Building the dam was one thing, but stopping the river was quite another. Nobody had ever put a dam across the Missouri before, and the stream fought like a bucking bronc to keep from being harnessed and driven through the four huge tunnels under the hills on the east bank. To do that, the channel through which the river was flowing (800 feet wide) had to be closed.

But when that was attempted, disaster threatened to delay the completion of the dam for many months.

The river, forever testing its strength against all obstacles, rose—in fact, 75 per cent higher than any stage attained since the beginning of the project. This high water weakened the toe gravel at the base of the narrow channel. Then, as the water went down, a slip in the embankment occurred, which threatened a pier of the railroad bridge crossing the river there. The threat increased swiftly. Something had to be done at once before that bridge went out— the only bridge across the Missouri within thirty miles.

The four tunnels through the hill were ready—except

THE GREAT DAM 283

for a 50-foot bank of earth built to keep the river from going into them during their construction. A big party had been planned to celebrate their completion.

Colonel Thomas B. Larkin, United States Engineers, in charge of construction, knew that he had no time for parties just then. He ordered those earth-plugs blasted out. This was done shortly after four o'clock in the morning, and within half an hour water was flowing steadily through the tunnels. Meanwhile Larkin rushed trainload after trainload of stone and gravel out along the shaky bridge to be dumped into the river channel. He had to raise the dump to a level where the entire current would be turned aside into the tunnels. And he had to be quick about it.

Other engineers shook their heads and assured him that the river was too strong to be checked by dumping at that point. The colonel, knowing his professional reputation was at stake, was appalled by the slowness with which the river rose—hardly more than one-tenth of an inch every quarter of an hour.

Unfortunately, because of the risk of derailment, time had to be taken out between trains while the gravel was swept from the tracks. A halt of only ten minutes caused the water level to sink again. It was a desperate crisis. But at last, with the 140th carload, the colonel saw with relief that the top of his new cofferdam was above water. Greatly cheered, he dumped more trainloads and so forced the Missouri into its new course underneath the hills.

The fill was completed with dredged-up earth, raising it to a level with the rest of the dam. The reservoir, or lakeside, face of the whole structure was later covered with eighteen inches of gravel, three feet of sizable boulders, and a top layer of quarried rocks weighing not less than one ton apiece. It was a great victory for Colonel Larkin, for his staff, and for the workmen, who formed a town with a population of five thousand souls. But the celebration scheduled for the opening of the tunnels was never held. The Missouri River had crashed the party, and its hosts had too much on their hands to take time out for a celebration that day.

Nonetheless, it was a unique triumph. For, as George Fitch has so well said, "It is a perpetual dissatisfaction with its bed that is the greatest peculiarity of the Missouri. It is harder to suit in the matter of beds than a traveling man. Time after time it has gotten out of its bed in the middle of the night, with no apparent provocation and has hunted up a new bed, all littered with forests, cornfields, brick houses, railroad ties, and telegraph poles. . . ." [2]

Though skeptics insist that the amount of water behind the Fort Peck Dam will never be sufficient to make the river below navigable, performance already contradicts them.

The dam has done its bit to win the war. Invasion barges, constructed in Kansas City yards 1,200 miles below, are floated on the first leg of their voyage to our enemies' beaches under their own power. This is made possible by sending water from the Fort Peck reservoir in a two-foot crest down to the Lower River.

But for all their enthusiasm for improvements on the Big Muddy, a great many of the people along the upper river are by no means wholehartedly behind the theory of unlimited government spending. One old ferryman made a few pithy and unquotable remarks about the money squandered in trying to make the Missouri navigable. He added, "This river is too big for engineers. If they want to have it navigated, they could. We rivermen know the river, and we got the flat-bottom boats to do it with."

CHAPTER XXV

Honeymoon Cruise

O N JULY 17, 1859, the first steamboat to reach "the mountains" arrived at Brulé Bottom, fifteen miles below Fort Benton, on her own power, a distance of 3,560 miles from salt water. No steamboat had ever gone so far from the sea before. The feat was as celebrated in those days as Lindbergh's transatlantic flight in ours. But the brave little *Chippewa* did not long survive its triumph. A year or two later, while upriver, some of the crew, sneaking into the hold to steal whisky, upset a candle and set fire to the boat. It burned to the water's edge at Disaster Bend.[1]

In 1866 the steamboat *Luella* reached Fort Union at the mouth of the Yellowstone and took on cargo for St. Louis. It was early in September and already frost was turning the leaves to color when a horde of miners from the diggings upriver reached the fort, eager to catch the last boat for the States.

Two hundred and thirty of them took passage on the *Luella*. Every man wore a buckskin money belt, and among them they had a million and a quarter dollars in gold dust— probably the richest cargo ever floated down the Missouri. The miners paid their passage in gold dust, and Captain Grant Marsh made a clear profit of $24,000 from the trip. The following summer, on a new boat, the *Ida Stockdale*, Marsh made a net profit for the owners of $42,594 for the round trip between St. Louis and Fort Benton—the largest profit made that summer by any boat on the river and twice the cost of the steamboat. It is hardly surprising that Marsh drew a salary of $1,200 a month.

Boats which went up so far lost no time on the return trip because of the shortness of the season of high water. In 1862 four boats reached Benton, in '66 thirty-one, in '67 thirty-nine. But by 1874, two years after the railway reached Bismarck, North Dakota, only six boats reached the mountains.

Before the railroads came in, boats in the mountain trade made handsome money. A cabin passage cost $300, deck passage $75 one-way, and freight brought 10 or 15 cents a pound. Several of these mountain boats earned as much as $40,000 in a single trip. Their names are still remembered on the upper river: the *Chippewa,* the *Benton,* the *Welcome,* the *Florence,* the *Only Chance,* the *Far West,* the *Western,* the *Mary MacDonald,* the *Fontanelle,* and the *Nellie Peck.* They were nearly all fast stern-wheelers of very light draft.

In 1880 the *Nellie Peck* carried a 19-year-old bride to Sioux City. There the young matron boarded the *General Meade,* of which her new husband, Charles P. Woolfolk, was captain. The boat, bound for Fort Benton, was piloted by T. J. Anderson and carried supplies to the traders and troops upriver. The captain of the boat had been a pilot as early as 1868, and he and Anderson took pleasure in teaching Mrs. Woolfolk to handle the wheel and ring the signal bells in the pilothouse. She was the only passenger aboard.

She occupied one of the two best staterooms aft, a spacious room with a regular bedstead, washstand, and closet built in. A door with a window in the upper half of it opened on the deck, which ran completely around the boat. Over the stern wheel, were the quarters of the stewardess. There the laundering was done and there one could take a bath. Nearby stood a battery of big rain barrels full of river water—and prickly pears, thrown in to clarify it.

The boat made about fifty miles a day, and the trip to Fort Benton afforded much of interest. Mrs. Woolfolk had a light rifle with which she engaged in target practice, and members of the crew hunted when game appeared on the banks.

At first there were plenty of fresh vegetables and fruit in the larder, but as time wore on and they mounted the stream, the crew and passenger depended upon dried apples, lemons from the bar, and potatoes.

Breakfast was served after eight o'clock, and at ten o'clock the cabin boy brought coffee to the pilot and the officers in the texas. Dinner was put on the table in the cabin at noon—a dinner which always included mashed or boiled potatoes. At four o'clock Mrs. Woolfolk used to join her husband and the pilot for tea or coffee, at which the main dish was always a delicious smoked buffalo tongue. Tongues and jerked beef were purchased at trading posts along the river. Supper was at five o'clock, and the cook is remembered for the excellent baked potatoes served with that evening meal. Fish frequently was on the menu; for whenever the boat tied up for the night, the crew would set a trotline. When there was no game, she says, fresh meat was provided from the cold-storage room. Bear and buffalo meat were too coarse for her; she liked elk meat best.

At Vermilion, South Dakota, the clerk of the boat went ashore, married his sweetheart, and brought his bride aboard. Those on the boat rolled out a tarpaulin for the happy couple, greeting them with whistles, bells, and horns in a good-humored shivaree. Afterward toasts were drunk in champagne, with which the bar was plentifully supplied.

On the way up, visitors often boarded the boat, and among these was the leonine Sioux, Chief Gall. Selling liquor to Indians was strictly against the law, and whenever the boat tied up at an Indian agency, the bar was closed and locked.

Gall sat on the guards—the railing around the deck—and tried to wheedle Mrs. Woolfolk into giving him a drink of firewater. She told him that she could not. But Gall, putting on a doleful face, declared that he was "heap sick" and needed it for medicinal purposes. Like other Indians in those days, he was carrying a small looking glass set in a decorated wooden frame. Mrs. Woolfolk had seen such mirrors flashing signals from the bluffs as the boat mounted the river, and finally offered to provide Gall with the cure he craved in exchange for his mirror. Gall readily agreed, and so, with the connivance of the steward, the bargain was effected.

The crew delighted in bringing wild animals aboard as pets for the captain's lady. She had several chipmunks which would sit on her shoulder or head and chatter while she fed them.

One morning she was awakened by a most unusual sound—a baby crying. At the time, the boat was tied up at a woodyard. Mrs. Woolfolk dressed hurriedly and went out on deck, supposing that one of the woodhawks had a wife and baby. The pitiful wailing continued; she thought perhaps the child was ill and needed attention.

But no such thing. The "baby" was a beaver kitten which the men had caught for her—a cute little thing with soft and silky fur. But its melancholy crying was too much for her—she made the men take it back where they had found it.

Fort Buford, a small military post, stood at the mouth of Yellowstone, and the *General Meade* had expected to make a landing there to discharge a shipment of flour for the Indians. But before the boat could land, the captain was warned that the Indians were angry because they had learned there was no coffee or tobacco on the boat for them. These supplies were following in the *Nellie Peck,* but the Indians were impatient and ugly.

Accordingly, the captain went on upriver several miles, anchored in midstream, and posted six guards to keep a sharp lookout against attack. He barricaded the pilothouse, putting up sheet iron on both sides and behind it. There they remained until it was safe to drop down and unload at the fort.

While the crew was busy, Mrs. Woolfolk went for a walk along the bank. Suddenly she saw a mountain lion. It was coming after her. She stopped, not knowing what to do. Fortunately, just then, a sergeant rode up. He escorted her back to the boat and offered the wholly unnecessary advice that she get aboard and *stay* there. Afterward she passed the time fishing from the deck, putting her catch into a bucket—which she emptied into the river when the day's sport was over.

Taking advantage of the high water, they hurried on to Fort Benton.

In those days the old adobe fort was unoccupied, but there were two large trading stores nearby. The *General Meade* tied up alongside the I. G. Baker warehouse and the crew and pilot went ashore. But when the hold was opened, they found that the merchandise nearest the hatch was that consigned to the other trader.

Mrs. Woolfolk was then in the pilothouse, where she could see all that was going on, when she heard the voice of the engineer calling up the speaking tube. He said, "Can you bring the boat around to the other landing?"

This challenge put her on her mettle. Now she was to be put to the test.

"I think so. I'll try," she called down.

Taking the wheel and ringing the bells to signal him, she backed the steamboat into the stream and swung it grandly up to the other landing without a mishap.

All that day and most of the night, the small crew was kept busy unloading.

Once more that summer, the *General Meade* went up to Fort Benton. When Mrs. Woolfolk went shopping at the trader's store, she found that nothing was sold for less than 50 cents. She paid that for a paper of pins—just for the novelty of it.

Afterward the boat swung downriver. Until September, the *General Meade* transported freight between the Coalbanks and Cow Island, now that the water above was too shallow for navigation.

Whenever the boat was moving, the principal amusement in the cabin was cardplaying. The long room was comfortably furnished—overhead oil lights, good carpets, and several armchairs. They played whist on the dining table.

The deck hands on the lower deck often gambled, and one night when one of them was a big winner, he told Mrs. Woolfolk that he wished his winnings to go to his wife at home. That same night he disappeared. It was supposed that someone had killed him for his money and thrown his body overboard. The clerk wanted to split his money among the officers, but the captain's wife saw that it went to his family in the States.

On the way downriver, they had passengers—an English lord and his valet from Canada, and a number of carpenters who had been building Fort Assiniboin. Every man was armed.

One day she saw a dark herd of about five hundred shaggy buffalo coming down to swim across the river. Everyone seized his gun and rushed to the guards to watch. The buffalo came on.

Suddenly the leaders of the herd saw the steamboat moving athwart their course. For a moment they stopped short. Then, turning downstream, the whole herd rushed along the bank, racing to get ahead of the boat. The people on the boat laughed, not knowing what to make of that. But when the bulls had got well ahead of the steamboat, the leaders turned and plunged into the river. Then an old-timer explained that buffalo were like that—on the Plains they always insisted on crossing the trail *ahead* of a wagon train. They had a strong dislike for being headed off.

Meanwhile, the boat, sliding down the swift current, soon reached the swimming animals, and the engineer had to stop the paddles. The men on deck shot several, and having butchered them on the bank, brought the meat and skins aboard.

Today Mrs. Woolfolk, now Mrs. M. I. Draper, is proud that she was the only woman ever to pilot a steamboat on the Missouri.

She knew many of the famous rivermen of those days. Once Captain Grant Marsh passed her boat tooting his whistle—only to run into a dead-end slough. Her boat went by in the main channel with one triumphant blast, while Marsh, on his deck, yanked down his cap in sheer disgust.

In her opinion the greatest steamboatmen on the river were Joseph and John La Barge.

CHAPTER XXVI

White Castles

At intervals above Fort Peck for hundreds of miles the Missouri River affords some of the most extraordinary scenery in North America. All early travelers devote page upon page and plate upon plate to describing and portraying the marvels along the stream. Even the matter-of-fact Lewis and Clark were carried away by what they saw, and the precise and scientific Prince Maximilian compares the scenery with that of parts of Switzerland, remarking that he was reminded "of the Mettenberg and the Eiger, in the Canton of Berne." The curious pinnacles, he says, looked "like the glacier des Bossons in the valley of Chamouny." Continually he refers to the valley of the Rhine.

The names given to landmarks along the stream, such as Citadel Bluff, Cathedral Rock, Eagle Rock, Haystack Butte, Burned Butte, give only a faint idea of the variety of striking formations to be seen there; naming them would exhaust the ingenuity of the most imaginative geographer. Of course, these strange structures must be seen from the river to be fully appreciated, and today few travelers pass that way. Throughout the course of its most beautiful and striking scenery, the Missouri is almost without a visitor. The Montana State Guide Book virtually ignores that scenic solitude. Photographers and painters neglect it—a fact as curious and incredible as the country itself.

Moreover, all this region, though now so lonely and neglected, is rich in historic sites—old campgrounds of the earliest explorers, old trading posts and forts, old steamboat landings, and Indian battlefields, such as that where Sitting Bull, singlehanded, killed the Crow chief, or the place where he saved the life of that enemy boy, Little Assiniboin, from his own bloodthirsty warriors, afterward adopting the lad as his "brother." To one who knows the Indian history of the plains, almost every mile along the stream suggests a story.

But now that the Fort Peck Dam has flooded the valley upriver for nearly two hundred miles, some of these historic sites and much fine scenery are under water. Until the dam breaks or the Missouri dries up there will be no more paintings made of that part of the stream.

Fortunately many of the finest things lie above this artificial flood.

One of the strangest and most striking sights along the river, painted by Karl Bodmer and enthusiastically described by his master, Prince Maximilian, stood not far below the mouth of the Musselshell.

One afternoon the prince was in his keelboat speeding upriver before a brisk favoring breeze. He sailed around the bend and saw, to his astonishment, two handsome castles crowning hills on the south bank. Apparently the roofs were of yellowish-red tiles, contrasting agreeably with the snow-white façades, both regularly marked by windows. The middle-aged, untidy little man peered through his spectacles at that amazing vision; he was quite ready to go ashore and pay a visit to those imposing buildings.

It was with some difficulty that his guides persuaded the prince that his "castles" were actually detached sections of a long horizontal stratum of white sandstone, running through the hills on that side. His "windows," they said, were simply perpendicular slits weathered out in the face of the stone and casting shadows, while the "roof" was not of tile but was merely the remnant of an upper, thinner stratum. Nothing daunted, the prince stoutly dubbed his rocks the White Castles, and so they have been called to this day.

But the White Castles are only the beginning of the amazing architectural scenery on the Upper Missouri. Above these it flows through Badlands of a most extraordinary and romantic appearance. Cliffs rise in many places nearly straight up from the water two or three hundred feet, formed of very white sandstone which weathers readily but lies protected under two or three horizontal layers of hard white freestone, covered by the dark, rich loam of the neighboring plain. This plain extends back a mile or so, where rocky cliffs again rise abruptly for perhaps another two hundred feet. Erosion has carved the soft sandstone into a thousand striking figures and majestic shapes. Everywhere one sees what appear to be ranges of large freestone buildings adorned with pale pilasters, long handsome galleries, pinnacles and parapets adorned with statuary, columns with pedestals and capitals entire standing upright or rising pyramidally one above the other until they terminate in a spire or finial.

These are varied by niches, alcoves, grottoes, and have the customary appearance of desolated magnificence.

To complete the illusion, great numbers of martins have hung their nests on every jutty, frieze, buttress, or coign of vantage and hover in flocks about these time-eaten towers. There seems no end to the visionary enchantment which surrounds the traveler—spires, domes, pallid ramparts, ruined balustrades, shrines, palaces, and terraced skyscrapers stand everywhere, buttressed and symmetrical, enriched with monuments and weathered statues thick as upon some old cathedral.

Passing between these silent, natural walls gives one a feeling of visiting the ruins of some ancient city. A man half expects the inhabitants to appear suddenly from shadowy doorways or to look out between the battlements topping some grim old keep. He has the feeling that he is trespassing, intruding. If the climate up there were a little milder, one might expect people to go and chisel out rooms in one of those neat sandstone piles and move right in. Walls and roof are there already. All that is lacking is an interior. There is nothing quite like all this in North America.

In our own time a poet has best celebrated the charms of this part of the river:

"Bad Lands? Rather the Land of Awe! . . .

"Rows of huge colonial mansions with pillared porticoes looked from their dizzy terraces across the stream to where soaring mosques and mystic domes of worship caught the sun. It was all like the visible dream of a master architect gone mad. Gaunt, sinister ruins of mediaeval castles sprawled down the slopes of unassailable summits. Grim brown towers, haughtily crenellated, scowled defiance on the unappearing foe. Titanic stools of stone dotted barren garden slopes, where surely gods had once strolled in that far time when the stars sang and the moon was young. Dark red walls of regularly laid stone—huge as that the Chinese flung before the advance of the Northern hordes—held imaginary empires asunder. Poised on a dizzy peak, Jove's eagle stared into the eye of the sun, and raised his wings for the flight deferred these many centuries. Kneeling face to face upon a lonesome summit, their hands clasped before them, their backs bent as with the burdens of the race, two women prayed the old, old, woman prayer. The snow-white ruins of a vast cathedral lay along the water's edge, and all about it was a hush of worship. And near it, arose the pointed pipes of a colossal organ—with the summer silence for music.

"With a lazy sail we drifted through this place of awe; and for once I had no regrets about that engine. The popping of the exhaust would have seemed sacrilegious in this holy quiet." [1]

The scene, however, is not always so solemn, for there are many grotesques and weird, amusing figures which suggest all manner of fantastic animal and vegetable creatures. It is a haunt fit for speckled moon-calves. Having shown what order and magnificence can do, it appears that Nature produced these gargoyles in a more lighthearted mood—the timeless jests of Mother Earth.

Yet these uninhabited "cities" were not always so. All over the astonishing castles, fortresses, altars, and proud towers mountain sheep cavorted in the good old days. They ran up cliffs like monkeys and marched across the face of steep bluffs like flies. Hunters were fond of their flesh and their skins, and the Indians loved to make spoons of their great curving horns. In fact, in some Indian languages the name given the animal is "spoon-horn." Spoons made of the horns of the rams are often beautiful shapely things, yellow, clouded like jade, or translucent—and quite capacious too, holding a quart or more. They are, in fact, ladles, and serve the purpose of a cup or bowl.

When chasing bighorns over rugged country, the hunter, gasping after his climb, might see the animal apparently leap into space from the edge of the cliff. Hurrying up to look down the dizzy height, he expected to see the ram lying dead at the bottom. But no such luck. There it went, safe and sound, bounding away among the ruined temples and stone toadstools. The hunter could only conclude that the animal had dived off and landed squarely on its massive horns; or, if he did not quite believe that himself, he usually tried to make the greenhorns believe it.

But there were animals which did dive off tall cliffs on this part of the Missouri. For in old times, before Indians had acquired guns or steel arrowheads, they knew how to lure or drive buffalo over these precipices so that whole herds crashed to their death in a moment.

This was accomplished by some daring young medicine man, swift of foot, disguised as a buffalo, with skin, horns, and ears complete. Placing himself between the herd and the precipice, he waited until his companions suddenly showed themselves between the herd and the prairie.

Instantly taking alarm, the animals would whirl round, uncertain which way to run. Then, seeing the disguised decoy speeding toward the river, they followed in mad flight, trusting to his leadership.

The buffalo cows quickly outdistanced the heavier bulls, racing after the sprinter at the speed of a fast horse. Only a fleet-footed man could hope to keep in their van, as they steadily gained upon him until they were close on his heels. But he, panting for breath, threw himself into some crevice under the brow of the cliff or into a narrow trench prepared in advance, just as the herd found itself on the brink of the bluff.

Too late the leaders saw the gulf below. Those behind pressed frantically upon them, conscious only of the hunters crowding on their flanks and rear. Those in front were violently shoved over and the others blindly followed—a cascade of living things falling hundreds of feet to the cruel banks of the river below.

Indians waiting along the river butchered the meat at their leisure, leaving what they could not use for the wolves swarming in to share the feast.

Lewis and Clark found the recent wreckage of such a slaughter. In fact they could not have failed to find it, so dreadful was the stench. Wolves, still feasting on the carcasses, were so gorged that a man actually killed one of them with his spontoon. The explorers dubbed the creek where they saw this Slaughter River.

But such bloody scenes are quickly forgotten as travelers mount the stream and reach the narrow gorge above the mouth of Arrow River—the valley of the Stone Walls.[2] There high ramparts of black rock rise from the water's edge. In the midst of the more fantastic forms of sculptured white sandstone, these vast ranges of Babylonian walls seem certainly to be the work of man, so regular are they. They tower up, each varying in thickness from the others, from one to twelve feet, and each evenly laid up, and as broad at the top as at the bottom. The stones are thick, black, and hard, intermixed and cemented with sand and talc.

"These stones are almost invariably regular parallelepipeds of unequal sizes in the wall, but equally deep and laid regularly in ranges over each other like bricks, each breaking and covering the interstice of the two on which it rests; but though the perpendicular interstice be destroyed, the horizontal one extends entirely through the whole work. The stones, too, are proportioned to the thickness of the wall in which they are employed, being largest in the thickest walls. The thinner walls are composed of a single depth of the parallelepiped, while the thicker ones consist of two or more depths.

"These walls pass the river at several places, rising from the water's edge much above the sandstone bluffs which they seem to penetrate; thence they cross in a straight line, on either side of the river, the plains over which they tower to the height of from ten to seventy feet, until they lose themselves in the second range of hills. Sometimes they run parallel in several ranges near to each other, sometimes intersect each other at right angles, and have the appearance of walls of ancient houses or gardens." [3]

If anyone wishes a walled garden or a fortress readymade, he can find them waiting for him at the Stone Walls. From the top of them a man may see the mountains on a fair day.

But the scenery on the river, however strange, unworldly, and beautiful it may be, is after all only a backdrop to the struggles and triumphs of mankind. Trouble, as the anthropologists say, is the natural habitat of man. Let us drop down to Cow Island. . . .

CHAPTER XXVII

Chief Joseph's Last Battle

During the spring season of high water, steamboats could go up the river to Fort Benton, seventy-five miles below the Great Falls of the Missouri. But in autumn, when the river was low, the head of navigation was Cow Island. Here freight was unloaded from steamboats to be carried on up to the fort in Mackinaw boats. Here, too, the year's supplies for the Royal Northwest Mounted Police were stored until they could be hauled by bull train up Cow Creek Canyon to the Canadian posts.

Just there was a ford known as Cow Island Crossing—the only good ford on this part of the river. Here on September 23, 1877, Chief Joseph and his handful of Nez Percés reached the river toward the end of their long running fight with the troops of the United States Army. Though he had only a few warriors and a large number of women and children, he had outmarched, outfought, and outwitted the best American officers who had been sent against him, inflicting heavy casualties upon them in one engagement after another, while trying in vain to find allies among the tribes along his march of nearly 1,500 miles. It had been one of the great retreats of history.

Having outsmarted Sturgis and Gibbon, Joseph pushed on, leaving the hostile Crows defeated behind him.

The Nez Percés had been friendly to the whites ever since Lewis and Clark first visited them. Their boast was that no member of their tribe had taken the life of a white man. They had peaceably parted with most of their lands in Oregon, retaining only the Wallowa Valley and a part of Idaho. But when gold was found in their country, the whites had swarmed in and the government proposed to move them to a new reservation at Fort Lapwai.

General O. O. Howard was sent to round up the Indians. He called a council, at which Joseph stated the Indian case in so pithy a manner that Howard had to fall back on violence to win the argument.

Joseph spoke as follows: "In treaty councils the commissioners have claimed that our country has been sold to the government. Suppose a white man should come to me and say, 'Joseph, I like your horses, and I want to buy them.' I say to him, 'No, my horses suit me, I will not sell them.' Then he goes to some neighbor and says to him, 'Joseph has some good horses. I want to buy them but he refuses to sell.' My neighbor answers, 'Pay me the money and I will sell you the horses.' The white man returns to me and says, 'Joseph, I have bought your horses and you must let me have them.' If we sold our lands to the government, that was the way they were bought."

But in spite of Joseph's logic, Howard arrested Too-hul-hul-suit, a prominent medicine man. This broke up the council.

To secure the release of that medicine man, Chief Joseph, in spite of the deathbed adjuration of his own father, at last, reluctantly, agreed to move. But his people, the southern portion of the tribe, took matters into their own hands and were already brandishing the scalps of white men before Joseph knew what was up. When the troops moved in, the Nez Percés gave ground at first until Chief White Bird made a flank attack and put the soldiers to rout. This was the first battle. After another fight on the Clearwater, the Indians decided to strike out for the British

Possessions to join Sitting Bull. Now they had reached the Missouri.

General Howard was deliberately dawdling on Joseph's trail, expecting rightly that the Indians, who could travel so much faster than the troops, would not hurry unnecessarily. Major Guido Ilges, with a company of the Seventh Infantry, waited above at Fort Benton for the Indians to strike the river; while Colonel Nelson A. Miles, marching overland from Fort Keogh on the Yellowstone and coming up the south bank of the Missouri, had just reached the mouth of the Mussellshell.

It was now six sleeps since the Nez Percés had fought the Crows and Bannocks.

That day the steamboat *Benton* had unloaded at Cow Island some fifty tons of freight and turned back downriver. The goods were guarded by Sergeant William Moelchert of the Seventh Infantry with twelve soldiers and four civilians. This guard dug in, throwing up a breastwork at some distance from the stores, so as to have a clear field of fire.

There were little more than one hundred warriors with Chief Joseph. But he had to cross the Missouri there if he was to join Sitting Bull in Canada, for this was the only shallow crossing anywhere near. When all his Indians had arrived at the ford, an advance guard of some twenty warriors crossed first, while the rest prepared to cover them from the south bank. The Indians waited for the soldiers to fire first.

Sergeant Moelchert, naturally enough, seems to have felt that he could hardly be expected to begin hostilities when so outnumbered. He held his fire.

The advance guard went over, then the rest of the warriors and their families, the pack train and the pony herd: all crossed the stream without opposition. All the women and a few of the men went on north of the river a few miles and made camp.

As they passed, the Indians helped themselves to the supplies. They had been on short rations for a long time, and now every family helped itself to sugar, coffee, rice, beans, hardtack, and flour. Some even carried off sides of bacon; which shows how hungry they were, since Indians had little taste for pork. The food was stacked up as high as a one-story house; there was plenty for everybody. That night the Nez Percés feasted. No trouble about cooking: they carried off pots and pans, tin cups and buckets, and everything else that pleased their fancy.

While this was going on, the warriors kept watch on the entrenched guard, and before long the guns began to pop. But it was so late in the afternoon by that time that the fight did not amount to much. It was soon dark. Finally the Indians set fire to the remaining stores and left the river for camp. Their hearts were good: nobody had been killed and they all had plenty to eat. Toward morning the last of the warriors left the fight.

Colonel Clendennin at Fort Benton soon after received the following letter:

"Colonel:
Chief Joseph is here, and says he will surrender for two hundred bags of sugar. I told him to surrender without the sugar. He took the sugar and will not surrender. What will I do?
Michael Foley." [1]

Two days later Major Ilges with thirty-six mounted men hit the trail up Cow Creek Canyon and had a brush with the Indians about noon. The warriors scattered along the high ground in his front. He soon found them such good shots that he led his small force in orderly retreat back to the rifle pits at Cow Island.

When Miles reached the mouth of the Mussellshell, he

CHIEF JOSEPH'S LAST BATTLE 303

met the steamboat *Benton* paddling down the river; but the captain, having left Cow Island before the Indians arrived, knew nothing of the trouble there. Miles, therefore, planned to follow up the south bank to the crossing. The steamboat stopped a few miles below to take on wood. Just then a Mackinaw boat came speeding down the current from Cow Island to tell Miles that the Indians had attacked and moved on.

Miles was frantic. If he had to march all the way up to Cow Island to cross, the Indians might be gone into Canada before he could reach them. He could never cross the broad, deep river where he was. That very day he had seen one of his scouts drowned when trying to cross. It would take forever to ferry his troops over in one Mackinaw. And he had let the steamboat go on downriver!

Miles could still see above the barren bluffs the tantalizing smoke of that steamboat moving down the river. There was no hope of overtaking it nor any recognized way of signaling a message. But he had a twelve-pounder fieldpiece. He snapped out an order: the gunner sent a shell crashing into the hills beyond the steamboat, kicking up the dust there. Still the boat moved on. A second shell followed the first. Then the captain, realizing that something had gone wrong, brought the boat back and, learning what was up, quickly ferried the command over to the north bank. If that Mackinaw had been delayed an hour, Chief Joseph might have made it to Canada, and Colonel Miles would have missed his promotion.

The Indians, having driven Ilges back and knowing nothing of Colonel Miles's command, remained carelessly in camp feasting and did not even scout their back trail. They traveled on leisurely through the Little Rockies and the Bear Paw Mountains, and had almost reached Milk River when Miles struck their trail from the east, crossed it and circled back to attack from the west.

It was easy to keep ahead of General Howard. The people were tired, tired of travel, tired of fighting, they needed rest and food. Scouts saw stampeding buffalo and warned Chief Looking Glass, now director of the march, that soldiers must have caused the stampede. But Looking Glass, perhaps a little puffed up by his new authority, perhaps put off guard by the many victories of his warriors, rode about the camp telling people not to be alarmed, that there was plenty of time. The disaster which followed was due to his delay.

The Indians had made their camp on Snake Creek, in a depression in that rough, barren country, while they hunted, killing buffalo and "making meat" for the winter. Snow was already falling on the heights, and a cold wind brought it to the Indians as rain. Some were already packing their animals for the march into Canada.

That same morning, not long after, some Indians out running buffalo saw the troops coming. One of them raced his horse to the bluff where the memorial monuments now stand and signaled with his blanket to the camp: "Enemies coming fast."

Chief Joseph's first thought was of the horses. While the people ran about in confusion, he called out his orders to save the animals. The whole camp was astir, like a disturbed anthill. Warriors were grabbing their cartridge belts and rifles, running to catch their horses, or scrambling up the bluffs to face the troops. They heard the thunder of the charging cavalry, saw the enemy sweeping toward them on three sides. Among them rode naked racing Sioux and Cheyenne scouts. That enraged the Nez Percés. Rifles began to crack. Elsewhere the Indian horses became frightened and ran away from their masters who were trying to catch them.

Miles had about six hundred men of the Seventh Cavalry, Second Cavalry, and Fifth Infantry, including Indian scouts. Chief Joseph had a little more than one hundred warriors. The Indian families outside the camp, with sixty warriors, immediately hit the trail to the north with the pack train while the remaining warriors manned the ridges.

It was a cold, windy morning with increasing snow flurries. Captain Hale, on being ordered to attack the camp, grumbled, "My God! Have I got to go out and be killed in such cold weather!" But to Miles that attack seemed his hour of glory. For twelve days he and his men had pushed hard on the trail of the Indians, fearing they would not catch them south of the Canadian boundary—and now he had found them. He gave the command to attack. The cavalry charged over hill and dale, hell-for-leather, head-on into the Indian field of fire. The warriors waited until the troops had moved up the slope to within a hundred yards. Then they cut down on them.

Other officers, who had tasted Nez Percé valor, would never have charged so rashly. The Nez Percés were good shots, and because they had never had too much ammunition, always shot to kill. They took aim and never fired in the air like rookies for the mere comfort of hearing the gun go off. Their Winchesters played havoc with the cavalry, killed or wounded most of the officers, emptied many saddles, and stopped Miles in his tracks.

"Miles and his staff passed several anxious minutes, uncertain as to what was happening. Then Lieutenant Erickson, wounded and covered with blood, rushed up to Miles and shouted in his face, 'I'm the only damned man of the Seventh Cavalry, wearing shoulder straps, who's alive!' He had seen every officer in his battalion fall, either dead or wounded.

"Miles ordered the Infantry to attack at once. They moved forward under heavy fire. The bugler was called to blow, 'deploy by the right flank.'

" 'I can't blow, sir,' he answered. 'I'm shot.' He lay

with a bullet in his spine. An order was called back to a sergeant in the rear to help him to cover.

"The bugler shouted: 'He can't do it, sir, he's dead!'

"Miles ordered the four-pound howitzer to a ridge northeast of the camp. As the team moved to execute the order, one rider and two of the mules drawing the piece were killed. It was abandoned with its muzzle pointed to the rear. . . ."[2]

The soldiers did not falter or turn back, but quickly took cover behind the blue line of their dead.

But the Second Cavalry on the right wing was a little out of the Indians' field of fire. It struck out for the pack train where Chief Joseph was, and he had to ride through that cavalry to reach his tent and get his rifle. His clothes were cut by bullets and his horse wounded, but he himself was not hurt.

The pack train, after a running fight, drove off the Second, which nevertheless captured some horses. Meanwhile there was another charge. In this attack the troops suffered casualties of 35 per cent.

Miles sent troops to cut the Indians off from their water supply, and there was fierce fighting hand-to-hand in the Indian camp, now entirely surrounded. But the Nez Percés soon forced the soldiers to pull out, leaving their wounded behind them. Still, the Nez Percés did not kill these helpless enemies left among their wretched tents; instead, they shared with them the very water the soldiers had been trying to take. Most of the Indians were Christians, and throughout their campaign fought with quite as much chivalry as the troops.

Miles was in a hurry. At any time, for all he knew, Sitting Bull and hordes of Sioux might swarm down from Canada to rescue the Nez Percés, or—what would be even worse—General Howard, his superior officer, or Miles's rival,

Sturgis, equally avid for promotion, might bring reinforcements and steal the glory of the victory.

Miles outnumbered the Indians five to one at least, but could not whip them. That day few of the redskins had been killed; and of these, in the confusion, four had been shot by mistake by their own tribesmen.

Of the whites, one battalion lost 53 killed and wounded out of 115 men. Captain Hale's "K" troop suffered a loss of 60 per cent.

But the Indians suffered also, losing several of their leaders, including Looking Glass. Had Looking Glass lived, it is possible that there would have been no surrender. By this time both sides had dug in; they sniped at each other during the cold snowy night.

Next morning a norther swept in, providing a perfect "smoke screen," had the Indians wished to take advantage of it and make their getaway; but Joseph was unwilling to abandon his wounded. Said he, "We had never heard of a wounded Indian recovering while in the hands of white men." The scouts with Howard's troops had killed and scalped all the old people and wounded Nez Percés left along the trail, and the massacre of Indian women and children at the so-called Battle of the Big Hole had left a bitter taste in Joseph's mouth.

Like any Indian captain, he probably felt a personal responsibility for these bitter losses and perhaps could not endure any more. For he felt he was to blame, as he himself later declared, for the two mistakes that had brought disaster to his forces: first, in not overruling Looking Glass and hurrying on to Canada; second, in not scouting his own back trail. So now, like many another conscience-stricken man, he went from bad mistakes to worse; and while trying to protect a few wounded men and lost children, faltered in his duty to his tribe.

Chief Joseph was a great man with a big heart, fit to

be the father of his people. But he seems to have lost his head when his own mistakes filled his ears with the cries of women and children. A war chief should be made of sterner stuff, able to put the welfare of all before the hardships of a few.

Of course, every man has his cracking point, and that very tenderness of Joseph toward his people which enabled him to lead them through the wilderness toward freedom was now the very motive which prevented him from going through with it. It was unfortunate for his people that he could not transfer his authority to some bolder spirit. Perhaps tribal custom made this impossible. Perhaps he felt his own responsibility too keenly. His wife, who surely knew him well, was wiser than he. She handed him his gun and said, "Fight!"

He still hoped, perhaps, that Sitting Bull would come tearing down from Canada to destroy Miles as he had wiped out Custer. Once they saw a galloping mass coming, but it turned out to be buffalo running from the blizzard.

Miles was in a hurry to arrange the surrender. Joseph hung back. Miles then, with a flag of truce, lured Joseph into his camp under a safe-conduct—and held him prisoner! But the Indians captured an officer and detained him as hostage until Miles was compelled to exchange prisoners.

The Indians claim that Joseph was handcuffed and rolled up in a blanket so that he could not move. It may be that that night of confinement in Miles's camp had something to do with cooling Chief Joseph's martial spirit. But when he returned to his people, he told how he had been treated and declared that the war *must* go on.

Miles demanded unconditional surrender. Joseph held out for terms which would permit his people to return to the lands they had been driven from.

The siege lasted five days. Miles shelled the Indians, but before anything came of it, Howard marched in on October

4th. The weather was clearing fast, making escape more difficult every hour. Joseph's brother was among the killed.

The leaders sat in council while the storm passed. Joseph talked to them about surrender, pointing out that it was Miles and Howard who first requested a truce and that he himself had never asked to quit fighting. Afterward he sent his message of capitulation:

"Tell General Howard I know his heart. What he told me before, I have in my heart. I am tired of fighting. Our chiefs are killed. Looking Glass is dead. Too-Hul-Hul-Suit is dead. The old men are all dead. It is the young men who say yes and no. He who led the young men is dead. It is cold and we have no blankets. The little children are freezing to death. My people, some of them, have run away to the hills and have no blankets, no food; no one knows where they are—perhaps freezing to death. I want to have time to look for my children and see how many I can find. Maybe I shall find them among the dead. Hear me, my chiefs. I am tired; my heart is sick and sad. From where the sun now stands I will fight no more forever." [3]

This was early in the afternoon of October 5th.

Joseph offered his reversed gun to Howard, but at a sign from the general, handed it to Colonel Miles. Other warriors did the same.

The Indians felt that they had *not* been captured. The battle had been a *draw,* and the white men had made *all* the concessions. General Miles, in fact, promised that he would send the Indians back to the land of their fathers. That was why they gave up their guns. It was the only way they could have been induced to give them up.

Of course, Joseph was not chief of *all* the Nez Percés. He could speak only for his own band. Still, some men of other bands surrendered.

But there were Indians in the party who had no mind to surrender. About a hundred of these, including about thirty warriors (two-thirds of whom were wounded), escaped during the night, crossed the Canadian border, and took refuge with Sitting Bull. The Sioux made them welcome.[4]

In spite of Colonel Miles's promise, the Nez Percés were not permitted to go back to their homes in the mountains, but were sent south to the Hot Place (Indian Territory, now Oklahoma), where many of them died of malaria, malnutrition, and homesickness. Eventually, however, Miles was able to have them sent back to their own country. Most of those in Canada also went home in time. But White Bird remained north of the line until he died.

That march of the Nez Percés was the most remarkable retreat in all American history—a long running fight against tremendous odds through the wildest country, yet victorious —until it ended in a draw. Joseph and his fellow chiefs deserve a high place among the best generals this country has produced.

Historians give Joseph and Miles all the glory of this battle. But we will remember the part played in the campaign by the Missouri River. But for that swift river, its steamboat and Mackinaw and Cow Island Crossing, the Nez Percés might have got safely to Canada and left Miles biting his ambitious nails in vain.

CHAPTER XXVIII

Gypsies of the Upper River

AFTER Chief Joseph and the troops had left the scene of his surrender for Fort Buford, White Bird and Sitting Bull led a mixed party of Nez Percés and Sioux down to the battlefields to dig up some ammunition which White Bird's warriors had buried there. Indians who were along have told me that on the way down they encountered half-breeds on Milk River.

In those days, when all that country was covered with buffalo, breeds from Canada regularly came south to hunt twice a year, leaving their parishes to "run meat," just like the Spanish buffalo hunters or ciboleros of New Mexico.

These so-called "Free People" of Canada were descended from Indian mothers and white men of all nationalities engaged in the fur trade. They were known as Bois Brulés by reason of their dusky skins, and also as Métis or Red River Half-breeds. They formed a distinct community—almost a tribe.

These amateur Indians were a cheerful, careless, happy lot of men with the great strength and fortitude of the voyageur, hospitable and decent, and had a courtesy and grace of manner which won the good will of all who met them. They took their families and their priests along on the hunt, and sometimes ventured far down the Missouri River to trade with the Sioux or "make meat." Many of them were of French ancestry, and these were Roman Catholics. Their language was a French patois enriched with words borrowed from Chippewa and Cree.

They traveled single-file in great caravans, and you could always hear them long before they came in sight. Their two-wheeled oxcarts, known as Red River carts, were made entirely of wood without a single piece of metal, and the wooden axles screamed and whined in dismal chorus. When buffalo tallow was plentiful, they greased the axles with it; at other times they let the axles squeal, making a racket you could hear a mile off.

As they approached your camp, the swarthy mounted hunters would fire volley after volley, emptying their guns, Indian-fashion, in proof of friendhip. On reaching their campground, they corralled their carts in a close circle, hub to hub, and pitched their skin tipis in a smaller circle some twenty feet within the carts. One such train comprised 824 carts, 104 tipis, 1,200 animals, and 1,300 persons, including some 300 hunters. When the horses and oxen had been watered, and the sun was setting, the men drove these animals into the corral. Each parish, or band, in the camp had its own captain. Each in turn posted a guard to keep watch throughout the hours of darkness.

The laws of the hunters were administered by their governor, or chief, and strictly enforced. These rules, like the costumes and customs of the people, were a blend of the Indian and the European. The regulations had to do mostly with hunting:

No one was to hunt or travel on Sunday. When going after buffalo, no one was to leave the group of hunters without permission, or run buffalo before the word was given. Penalties for offenses against these rules mounted from the destruction of the offender's saddle or coat to a severe beating.

If anyone in the camp was convicted of stealing anything, however valueless, he was brought to the middle of the camp circle, where the crier shouted his name three times, at every repetition adding the word "thief." But such penalties seldom had to be exacted.

When the buffalo scouts returned to announce that they had located a herd, the governor took command. Every hunter saddled up, threw off unnecessary clothing, tightened his belt or sash, and all prepared to move off in a body. If the herd was at a considerable distance, the men rode pack horses and led their trained buffalo ponies, sparing these for the race ahead.

Once near the herd, they all halted behind the last ridge out of sight of the buffalo, until the chief gave the word to charge. Then over the hill they went, pell-mell—and away went the buffalo ahead of them. Plunging through clouds of dust on the rear or flank of the fleeing herd, the buffalo pony would lay his master alongside a plunging cow. Then the rider, without taking aim, lowered the muzzle and fired at close range into the heart or spine of his prey. As fast as he fired, he would pull the plug of his powderhorn with his teeth, hastily pour in a fresh charge, set the muzzle of his gun to his mouth, drop in a wet bullet, settle the charge by pounding the stock of his gun on the pommel of his saddle, and then fire again. In this way he dropped one after another of the shaggy cows.

Within half an hour the plain was covered with dead or dying bison, while over the ridge ahead rolled a few old bulls not worth killing. The hunters then dismounted, and leading their heaving, lathered horses, went back to identify the animals each had shot.

Meanwhile the women, driving the screeching carts, came up to join the men in skinning and butchering the carcasses.

For several days thereafter the women were busy in camp making meat.

This was cut into long strips or thin flakes and hung

out to dry in the sun or over a fire. When dried, certain parts were folded up and packed in bundles weighing about sixty pounds. The rest of the meat was piled on the raw hides pegged to dry on the ground, and now stiff as boards.

The dry meat was brittle and easily beaten to a powder on this primitive threshing floor. Meanwhile the fat was cut and melted in iron kettles.

Afterward they poured the hot tallow upon the pulverized meat and stirred the mess with spades until it was well mixed and thoroughly saturated. This mixture, to which dried pounded plums or chokecherries were sometimes added, was then packed in rawhide sacks covered with hair, and sewed up snugly with sinews. As the rawhide dried and contracted, the pemmican was squeezed together into a hard, compact mass.

Such a sack contained a balanced diet and was the staple ration of the plains. Each sack was known as a "bull" (taureau). If the fat used was the fat from the udder of the cow, the sack was called a "fine bull" (taureau fin). A sack containing dried fruit as well, they called "taureau à graines." The first was good, the second tasty, and the third delicious. One cow could provide only half a bull and about three-quarters of a 60-pound bale of meat. The meat of eight or ten cows made a load for one cart.

The marrow from the bones was melted and preserved in the bladders of the animals. One bladder held the marrow-fat of two cows.

These Canadian breeds made heavy inroads on the buffalo herds, and therefore sometimes had to fight the Sioux. We have record [1] of the kill of one small party of 55 hunters who carried away the meat of 1,776 cows which (Belcourt says) "formed 228 *taureaux*, 1213 bales of dried meat, 166 *boskoyas* or sacks of fat, each of them weighing 200 pounds, and 556 bladders of the marrow, of 12 pounds each; the entire amount, estimated at the lowest rate, being worth a little more than 17,000 pounds sterling. The expenses of the trip and the wages of the employees amounting only to about 200 pounds sterling, there remains 1,500 pounds sterling, earned by 55 hunters in the space of less than two months, counting from the day of departure to the day of return."

After the hunt all was gaiety in the camp, singing and fiddling and dancing on the stiff pegged-down hides under the stars. It made a picturesque festival—the men, in their blue Hudson's Bay blanket coats, or capotes, adorned with big brass buttons, belted with gay sashes and wearing leggins and moccasins; the women in bright calicoes and colorful blankets.

The independence of these colorful and pleasant people ended suddenly when they were led to rebel against the British by Louis Riel, and they came no more to hunt south of the boundary line. But their memory is still green on the Upper Missouri, and their unique cart is paired with the bullboat as one of the two vehicles most characteristic of the Big Muddy.

CHAPTER XXIX

Great Falls

AFTER leaving the canyon of the Stone Walls, going up the river into the Blackfoot country, Lewis and Clark reached the mouth of Maria's River, named by Captain Lewis after his cousin, Miss Maria Wood, though, as he says, the "hue of the waters of this turbulent and troubled stream but illy comport with the pure celestial virtues and amiable qualifications of that lovely fair one." At this point the explorers were perplexed, not knowing which stream to follow—Maria's River or the Missouri proper. The only clue they had was that there were great falls on the Missouri. They therefore divided the party and explored until they could be sure. Clark then proceeded with the boats up the main stream while Lewis and four men walked overland.

As Lewis advanced he heard the agreeable sound of falling water and saw spray rising high above the plain, borne on the wind like smoke. The noise soon became tremendous, and, after marching seven miles, he reached the Great Fall at noon and seated himself on some rocks below to enjoy the wonderful spectacle which he was perhaps the first American to see.

Lewis attempted to describe the falls, which he says filled him "with pleasure and astonishment," but he was so "disgusted with the imperfect idea" his description gave that he "determined to draw my pen across it and begin again." Yet the record made by these explorers of the falls was so accurate that modern engineers have declared their errors of measurement to be within one foot of the present figures.

The falls are the principal cascades of a series that occupy the river bed for a distance of more than ten miles. The portage around these rapids is eighteen miles long. The five main falls, as measured by Captain Clark, are as follows:

Black Eagle Fall .	26 feet	5 inches
Colter Fall . . .	6 feet	7 inches
Rainbow Fall . .	47 feet	8 inches
Crooked Fall . . .	19 feet	
Great Fall . . .	87 feet	¾ inches

The Black Eagle Falls were so named from the fact that "just below the falls is a little island in the middle of the river well covered with timber. On a cottonwood tree an eagle had fixed its nest, and seemed the undisputed mistress of a spot, to contest whose dominion neither man nor beast would venture across the gulfs that surround it, and which is further secured by the mist rising from the falls." There the nest certainly remained for fifty-five years after. Indeed, it is reported that the bird, or at any rate an eagle, made its home there as late as 1872.

Soon after Captain Meriwether Lewis, "the sublime dandy," reached the falls, a series of adventures befell him which would have delighted his friends in Washington, D.C.

His hunters were all busy killing buffalo and jerking meat, fishing and cooking; but for Lewis, the falls meant a whole series of hairbreadth escapes. He climbed a hill and then descended toward the Sun River. All about were hundreds of buffalo. He shot one and watched to see it tumble so intently that he forgot to reload. Suddenly he spied a large brown bear creeping upon him only twenty yards away. Automatically Lewis raised his rifle and, at the same time, remembered that it was not loaded. He knew that he had to take cover, and there was not a bush or a tree within three hundred yards. The bank of the river itself was not more than three feet high; he could not hide himself under that.

The captain attempted to retreat at a walk, increasing his pace as the bear advanced. But the moment he turned, the bear, with open mouth, rushed upon him.

For eighty yards or so the captain ran as fast as the bear. Then, finding it gaining rapidly upon him, he decided to take to the water. If the bear had to swim to attack him, he would have some chance, he thought, to defend himself.

With him he carried his spontoon or military half-pike, a sort of bayonet on a stick; with that he hoped he might cope with the animal.

Lewis plunged into the stream and waded out some twenty feet, until the water reached his waist. Then he turned and waited for the beast to attack.

At the water's edge the bear hesitated, seemed to lose its nerve. Then, suddenly, it wheeled, and ran off.

Lewis, dripping, walked on, congratulating himself on his escape, and resolved never again to allow his rifle to be a moment unloaded.

He went on exploring the river, when suddenly he met a brown and yellow animal standing near its burrow. At first he thought it a wolf, but as he came near, it crouched, as if to spring upon him. The captain fired, and the creature fled into its hole and disappeared. From its tracks and general appearance, Lewis supposed the animal to be "of the tiger kind."

He had hardly disposed of this danger when, out of a nearby herd, three buffalo bulls charged him at full speed.

This time the captain disdained to fly. The bulls came thundering on to within a hundred yards. There they stopped, stared at him for a while, whirled, and ran back to the herd.

It was now getting dark, but he kept on, reflecting on his strange adventures and the sights he had seen, "which crowded on his mind so rapidly that he should have been inclined to believe it all enchantment if the thorns of the prickly pear piercing his feet did not dispel at every moment the illusion." Reaching camp he found himself very tired; he ate heartily, and slept well.

He might not have slept quite as well had he known that when he waked he would find "a large rattlesnake coiled on the trunk of a tree under which he had been sleeping."

Having made the Louisiana Purchase, President Jefferson had sent Captain Lewis to see what he had bought. That day the captain found out. Bear, buffalo, "tiger," and rattlesnake—the Missouri had shown him just about everything the country afforded. And all in one day!

Somehow adventures on that part of the river always seem to have a happy ending.

The town of Great Falls, Montana, derives its prosperity mainly from the hydroelectric plant there. It is a pleasant town opposite the mouth of Sun River and looks out on four ranges of mountains. Nearby are the Giant Spring, good fishing, and plenty of fine scenery.

But to most people, Great Falls is simply the home town of Charlie Russell, the pioneer "cowboy artist" of Montana. There, in the Mint, a frontier saloon preserved intact, one may see a large collection of his paintings, while his log cabin studio, purchased after his death, is Montana's most important museum of art.

Charlie Russell belonged to the Missouri River throughout his life as few artists have. Born in St. Louis in 1865, he very early felt a passion for the West, and saved up his money to run away and kill Injuns. When his mother found and deposited his money in the bank, he ran away anyhow, trusting to his luck. He starved three days before going home. Sent east to school, he neglected his studies to draw pictures and read western history. So, finally, the family gave in and sent him to Helena. There he fell in with bull-whackers and Indians and hit the trail with his boss for a Judith Basin sheep ranch.

But Charlie was no sheepherder. He soon threw in with a trapper and spent two years on the trapline and hunting trail, packing in meat to the settlers and trading his furs at Fort Benton. Holed up during the winter, he began to paint and sketch. When his father sent him money to return home, he sent the money back and stayed on the Missouri.

Before long he became a cowboy. Charlie was always modest about his qualifications—in contrast to some "cowboy artists" who perhaps know nothing of range life except what they have seen on a dude ranch. "I was neither a good roper nor rider. I was a night wrangler. How good I was I will leave it for the people I worked for to say. . . . I worked for the big outfits and always held my job. . . . I haven't been too bad nor too good to get along with. . . . I believe in luck and have had lots of it."

In the winter of 1886-87—that terrible winter which wiped out so many thousand head of cattle on the plains—that winter when the frost was so severe that the horns of cattle burst with the cold—Charlie Russell came to sudden fame. At the time he was on the ranch. His employer, Louie Kaufman, alarmed by reports of stock losses, wrote to Russell, asking about the condition of his herds. Instead of writing a letter in reply, Charlie got out his water colors and painted a melancholy picture of a starving Bar-R cow knee-deep in snow, looking at some waiting coyotes. Charlie entitled the picture "Waiting for a Chinook, or The Last of the Five Thousand."

This picture told the story so much more eloquently and vividly than a letter could have done that it immediately called the artist to public attention.

From then on he painted and sketched and modeled

until his works were more in demand than those of almost any other contemporary American artist. The Mint now does a thriving business selling color prints of his work. But the man himself was loved even more than the fine record he made of that Upper Missouri country he loved and painted so well. He himself said of his work, "To have talent is no credit to its owner. . . . Any man that can make a living doing what he likes is lucky, and I am that." When his work was praised, Charlie merely said, "Nature has been my teacher; I will leave it to you whether she was a good one or not." [1]

Charlie Russell is *the* painter of the Upper Missouri. Most lovers of the Old West bracket him with Frederic Remington and Charles Schreyvogel, and he is by no means the least of the three.

His bronzes are, if possible, even more authentic and vital than his pictures.

But Charlie did not stop with the graphic arts. He published two illustrated books, both classics of the West, both having that easy, relaxed awareness so characteristic of the plainsman. These are, of course, his *Trails Plowed Under* and *Good Medicine*.

CHAPTER XXX

The Gates of the Mountains

ABOVE Great Falls, the river flows northeast, through the country of the Blackfeet, but skirting the old-time range of the less warlike Flatheads.

One of the most touching tales of the Missouri River is that of the delegation of four Flathead or Nez Percé Indians who, in the spring of 1831, started from their home near the Bitter Root Mountains and came down to St. Louis. Their purpose was to find Christian missionaries to bring the true faith to their people.

It appears that their people had learned something of the Christian religion from Canadian Indians visiting their camps. They had come all that long, dangerous journey to seek the light of which they had heard such good reports. Two of them died in St. Louis and were buried there in the cemetery of the parish in October and November of that year. No one in St. Louis could understand their language.

Finally, however, they found their way to the office of General William Clark, the explorer, who was then Indian agent. Through him their message was made known to a Protestant, who called the matter to the attention of his church. The two survivors traveled with the artist, George Catlin (so he says) some two thousand miles. He painted two "portraits," and gives their names as Rabbit-Skin-Leggins and No-Horns-On-His-Head. General Clark described the two as Nez Percés.

The tribe of these seekers has been much disputed, some

calling them Flatheads, some Nez Percés, while others think it possible that they were of mixed origin or formed a joint delegation.

The controversy is in itself amusing. For it appears that the Canadians who first brought word of Christianity to the Indians in the Rockies described missionaries as Black Robes (Roman Catholic priests) and that the delegation naturally came seeking Black Robes to help their people. However, the Protestants first got wind of this Macedonian cry and promptly sent out missionaries of their own before the Catholics took any steps in the matter. In 1834 the Methodist Episcopal Church sent out the Reverend Jason Lee and his nephew, the Reverend Daniel Lee, with some laymen, to found a mission among the Flatheads. These men, however, passed on to Oregon to preach the gospel on the Willamette. Soon after, the Reverend Samuel Parker and Dr. Marcus Whitman met the Flatheads and Nez Percés. But, as the Jesuit Father L. B. Palladino gleefully relates, in his instructive book,[1] the Flatheads would have none of them. For, like the Lees, they had wives, wore no black gowns, carried no cross, and did not perform the ceremonies or say the Big Prayer of which the Indians had been told. In short, while the Nez Percés were willing enough to listen to Protestant missionaries, the Flatheads insisted to a man on the genuine Black Robes.

The message of these four Indians, and their long journey involving the death of two of them, made a powerful appeal to the missionary spirit of America and caused a great many missions to be established throughout the West. Many an earnest worker went to labor in the vineyard along the Missouri.

Of these early missionaries, the Jesuit Father Pierre Jean de Smet appears to have made the longest journeys through the Valley, and to have received most notice from historians. He loved the wilderness and the Indians, and the Indians loved him. In 1837 he went by boat from St. Louis to Westport to join the annual expedition of the American Fur Company heading for rendezvous on Green River. From that date on he spent years traveling through the wilderness and became so influential among the Indians that he was able to serve in a political capacity when no one else could act. In 1868 he ventured with only one other white man in his company from Fort Rice to Sitting Bull's hostile camp on Powder River, and there arranged for the treaty of peace later held at Fort Rice and known to history as the "Treaty of Laramie."

On that occasion he gave Sitting Bull a small crucifix, now in my possession; but, though Sitting Bull respected, admired and loved the great Blackrobe and followed his advice as to making peace, the chief would not submit to baptism or abandon his gods even for Father de Smet.

The esteem in which the good father was held by the Sioux is proved by a saying common among old-timers that Father de Smet and General William S. Harney were "the only white men they ever knew who talked sense and told the truth."

Father de Smet celebrated the first Catholic mass in Montana for the Flatheads in 1840. He also worked among the Blackfeet.

One day he preached to a crowd of these Indians through an interpreter. When he had finished, one of the chiefs, all painted and feathered and dressed in buckskins, came down and shook hands with him. He amazed De Smet by speaking very good English, and told the priest that he had employed a very poor interpreter. " 'These people,' said the Blackfoot chief, 'are deeply interested in what you have preached to them, but your interpreter did not put it before them in the proper way.'

"But you, please, sir, where did you learn English?" asked Father De Smet in amazement.

'Faith! In Ireland,' replied the Blackfoot chief." [2]

The Irishman went on to tell how he had come west, and there became too fond of the bottle. But an old friend, an Indian trader, in order to save him from drink, had taken him along on one of his expeditions. There he met the Indians, and took to their ways like a duck to water. One day when their enemies attacked, his Irish valor had so impressed them that they had made him a chief.

" 'After that,' said he, 'I married a squaw as well as I could, where no sight of a priest was to be had, and I have five papooses whom I have baptized myself, as well as I knew how. But I'd like your Reverence to do it all over for me and do it right this time.' " [3]

So Father de Smet baptized the Irish chief's papooses. But before long there were plenty of men in that country who needed moral instruction far more than ever the Indians did. On both sides of the river above the bend, gold was found.

The first strike was made in the fifties. Men who had learned mining in California, drifting back from the coast, heard that a half-breed had found colors in western Montana on Gold Creek. Miners moved in from Colorado, and in 1862 on Grasshopper Creek, two of them struck pay dirt. Immediately a camp called Bannack sprang up there. A little later richer strikes caused a rush to Alder Gulch, Virginia City, and Last Chance Gulch.

Confederate Gulch outdid all the rest. It was named for some Confederate soldiers taken prisoner in Missouri by the Union forces and sent upriver. As the Civil War drew to a close, two of these fell to prospecting, operating in the Big Belt Mountains east of the river. There they soon found gold enough. Prospectors swarmed in—some of them experienced miners; others utter greenhorns.

One of these latter, they say, was so green that he went

up to an old-timer busy with his pan and asked him to point out a better place to dig.

The shaggy miner straightened up and stared for a moment at his naïve questioner. Then he looked around—to pick out the least likely spot in sight. He spat and pointed up the creek. "Try that bar yonder. Who knows? Maybe you will find something."

Taking the advice in good faith, the greenhorn followed the miner's directions, staked his claim, Montana Bar, and went to work. The bar covered about two acres. It has been called "the richest acre of gold-bearing ground ever discovered in the world." Up to that time, yields thereabouts had never gone above $180 to the pan, but that greenhorn found gravel yielding over a thousand dollars to the pan! Panfuls of clean gold were taken out of Montana Bar at a single cleanup, weighing some seven hundred pounds and worth $114,800. A single shipment by wagon to Fort Benton—over two tons of gold—was valued at more than a million dollars. Within four years the gulch produced ten million dollars in dust and nuggets.

Numbers of nuggets were found in the region worth from $100 to $1,800 each; several were valued at more than $3,000 each. Some of the ore was so rich that it was shipped by wagon to Fort Benton, by boat to sea, and by ship to the British Isles—and still made a handsome profit.

Wild booms followed those first discoveries, and all sorts of greedy and vicious characters swarmed into the diggings to share in the sudden prosperity of the miners. Dance halls and saloons were everywhere. Every camp was wide open. Road agents and stickup men abounded, and before long these organized themselves into a gang.

Distances were great; and the government in territorial days was far away. Finding that the local law-enforcement officers were helpless or even in cahoots with the criminals, decent citizens of Montana organized vigilance committees for the trial and punishment of offenders. These committees were secret, swift, and sure in dealing with the murderers and robbers all about them, and made it their practice rather to make examples of ringleaders than to hang small-time bandits and killers. They strung up Henry Plummer and the notorious J. A. Slade, whose reputation as a bad man was so great that he supposed himself above the law.

But these thrilling stories of crime and punishment were seldom enacted on the Missouri River. Though the vigilantes captured, tried, and hanged a number of murderers at Helena and at Diamond City, these were practically all small-time, sordid criminals unworthy of our memory.

Most of the gold camps were not directly upon the banks of the Missouri River. One of the nearest was that in Last Chance Gulch, now the main street of the capital city of Montana. As the town acquired size and dignity, a meeting was held to adopt a more suitable name. The choice was Helena—a name borrowed from somebody's home town in another state. He pronounced it with the accent on the second syllable.

But the miners and the mule skinners, the road agents and the vigilantes would not have it that way. They claimed that the town was named after Helen of Troy, and their sentiments, which have prevailed, were expressed in the familiar jingle:

> Helen-a; after a darling, dizzy dame,
> Of much beauty but spotted fame:
> In pronouncing the name, understand me well,
> Strong emphasis should be laid on Hel.[4]

The days of the vigilantes and their victims are over. Where they operated or danced at the end of a strangling rope is now a region of peaceful farms and hospitable dude ranches. One might think the drama of the region had departed.

But the Missouri River is not one to let you down. From the mouth upwards it has offered many scenes of interest and strange beauty. But in its upper reaches it outdoes all these, bringing our journey to a climax of glory and grandeur worthy of the mightiest stream in all America. Not far below Helena, as one ascends the river, one enters a winding canyon of gigantic rocks, surpassing everything below. And from this point onward to its headwaters, the stream flows through a perfect welter of mountains. The great gorge was cut by the Missouri River straight *through* the Big Belt Mountains, and though many have tried their hands at describing it, the best description is still that first account penned by Captain Meriwether Lewis.

As he approached the gorge, and while still miles away, he was struck with wonder at the vista opening before him:

"A mile and a half beyond . . . the rocks approach the river on both sides, forming a most sublime and extraordinary spectacle. For five and three quarter miles these rocks rise perpendicularly from the water's edge to the height of nearly twelve hundred feet. They are composed of a black granite near its base, but from its lighter colour above and from the fragments we suppose the upper part to be flint of a yellowish brown and cream colour. Nothing can be imagined more tremendous than the frowning darkness of these rocks, which project over the river and menace us with destruction. The river, of one hundred and fifty yards in width, seems to have forced its channel down this solid mass, but so reluctantly has it given way that during the whole distance the water is very deep even at the edges, and for the first three miles there is not a spot except one of a few yards, in which a man could stand between the water and the towering perpendicular of the mountain: the convulsion of the passage must have been terrible, since at its outlet there are vast columns of rock torn from the mountain which are strewed on both sides of the river, the trophies as it were of the victory. Several fine springs burst out from the chasms of the rock, and contribute to increase the river, which has now a strong current, but very fortunately we are able to overcome it with our oars, since it would be impossible to use either the cord or the pole. We were obliged to go on some time after dark, not being able to find a spot large enough to encamp on, but at length about two miles above a small island in the middle of the river we met with a spot on the left side, where we procured plenty of lightwood and pitchpine. This extraordinary range of rocks we called the Gates of the Rocky Mountains."

CHAPTER XXXI

Three Forks

Above the Gates of the Mountains all the way up to Three Forks, the river flows clear, swift, and cold over a gravel bed—a very different stream from that which meanders through the muddy flood plain so far below. On either bank the whole country here lies lovely and unspoiled. And so we finally reach that broad valley or park walled in by mountains, where three clear rivers join within a mile to form the Missouri.

When Lewis and Clark reached Three Forks, Lewis climbed a limestone bluff, since known as Fort Rock, in order to overlook the union of these three rivers and decide which was the main stream. But that was a great puzzle, since each of the rivers was about ninety yards wide and all were so much alike that nobody could surely say that any one was the main stream. After observing the country, Captain Lewis, still unable to make up his mind, very sensibly "descended to breakfast."

There is virtue in a decision made after eating. The explorers finally voted to give each stream a name of its own —the Gallatin, Madison, and Jefferson—appropriately attaching the president's name to the westernmost, which appeared to be slightly the largest. When they returned to Washington and reported to the president, Jefferson remarked dryly that they had been a long time finding a stream to bear his name; to which one of the explorers diplomatically replied that they had been a long time finding one worthy to bear it.

Here in the midst of the richest fur country in the West, the Missouri Fur Company built a fort in 1810. But the Indians would not have it, and Lisa's men had to leave. There were too many hostiles about.

For Three Forks was then the crossroads of the wilderness. Indian trails ran in all directions, and down these trails the warriors came to many a bloody fray—Blackfeet from the north, Crows from the east, Snakes from the south, Nez Percés and Flatheads from the west. Here John Colter, discoverer of Yellowstone Park, helped by Crows and Flatheads, fought and defeated a horde of Blackfeet. Not long after, the Blackfeet caught him, disarmed him, stripped him, and made him run his desperate, naked race with death.

Here, too, in 1864, some settlers coming overland from Missouri built Gallatin City, which they supposed would be the head of navigation. But when they learned of the Great Falls below, they promptly abandoned their city. Today, not far away, the pleasant little town of Three Forks thrives.

The lucid streams which form the Missouri head far up in the Rockies, hundreds of miles above. Their waters join here at an altitude of more than four thousand feet to begin the long journey of more than twenty-five hundred miles to the mouth of the Missouri. It is fitting that this mighty river of the Plains should leap full-grown from the mountains in the midst of this "green, extensive, meadow of fine grass," this "handsome, level plain"—a place not only beautiful, but crowded with memories of old far-off things, and battles long ago.

Sacajawea rejoiced to find herself here. And so may we, as we consider the wonderful past and glorious future of this strange and passionate stream, and remember the steady verdict of Walt Whitman: [1]

THE MISSOURI

Others may praise what they like;
But I, from the banks of the running Missouri,
  praise nothing, in art, or aught else,
'Till it has breathed well the atmosphere of this
  river—also the western prairie-scent,
And fully exudes it again.

Notes

Notes

Chapter I

THE FATHER OF NAVIGATION

[1] *Nebraska History Magazine.* Missouri River Number (Illustrated). Vol. VIII, No. 1, p. 5. "The Missouri River Region as Seen by the First White Explorers," Addison E. Sheldon.

[2] "A Race on the Missouri," Charles D. Stewart, *The Century Magazine*, Vol. LXXIII, No. 4, February, 1907, p. 588.

[3] "The Missouri River, Its Habits and Eccentricities Described by a Personal Friend," George Fitch, *The American Magazine*. Vol. LXIII, No. 6, April, 1907, pp. 637-638.

[4] Sioux City (Ia.) *Register*, 1868.

[5] Such a craft is a great rarity today. That in the museum at Sioux City is a fine specimen, recently discovered in the sands of the river bed. It was carefully made of a seasoned walnut log, and though well-worn, is nearly as good as new. Over-all length: 12 feet, 11 inches; overall beam: 24½ inches; inside beam, 22 inches; inside depth, 14 inches. The sides are slightly more than one inch thick; the bottom 2½ inches thick. The bottom of the canoe has no keel, but is flat over a space ten feet long by fifteen inches wide. The prow and stern rise about 14 inches above the lowest part of the gunwale. The whole weighs 225 pounds. The canoe has no bulkheads, but on the floor toward either end a rib was left high enough to support a plank between them and so keep the cargo above any water that might be shipped. From the marks made by the workman's ax or hatchet, it appears that the tool had a narrow blade. Apparently the tree from which this canoe was hewn was well over 3 feet in diameter, for only three knots appear, and the canoe is still true. In the prow there is a well-worn hole for the cordelle. These measurements are offered here through the courtesy of Mr. W. R. Felton.

[6] For reproduction in color, see *Fortune*, February, 1940, p. 97. For Kurz, see Plate 25, Bureau of American Ethnology, Bulletin 115.

Chapter II
MISSOURI MARATHON

[1] Republished in *Early Western Travels, 1748-1846*, Edited by Reuben Gold Thwaites, LL.D. Vol. V, *Bradbury's Travels in the Interior of America, 1809-1811*, pp. 39-40. Cleveland, Ohio, 1904. The translation is Bradbury's.

Chapter III
STEAMBOAT 'ROUND THE BEND

[1] See "Pleasure Trips on the River Part of Brightest Memories," Kansas City *Star*, June 30, 1935.
[2] *Conquest of the Missouri*, Joseph Mills Hanson, Chicago, 1909.

Chapter IV
LONG'S DRAGON

[1] *Nebraska History Magazine*. Missouri River Number (Illustrated). Vol. VIII, No. 1, p. 32. "Life on the Missouri," Capt. Phil E. Chappell.
[2] "The Missouri River, Its Habits and Eccentricities Described by a Personal Friend," George Fitch, *The American Magazine*, Vol. LXIII, No. 6, April, 1907, p. 640.
[3] Publications of the Nebraska State Historical Society, Vol. XX, p. 349.
[4] Publications of the Nebraska State Historical Society, Vol. XX, p. 350.

Chapter V
PRINCESS OF THE MISSOURI

[1] His achievements have been recorded, from the incomplete records of his work, by Baron Marc de Villiers in *La Découverte du Missouri et l'Histoire du Fort d'Orléans (1673-1728)*, Paris, 1925. Only two of de Bourgmond's own manuscripts remain: first, *The Exact Description of Louisiana*; second, *Route to Follow to*

NOTES   337

Mount the Missouri, etc. The latter gives a day-by-day story of the movements of his party.
² See *The Lions of the Lord,* Harry Leon Wilson, Boston, 1903, Chapter IX, pp. 117-121.

CHAPTER VI

THE BIG MUDDY

¹ This question has been discussed at length in *The Missouri River and Its Utmost Source,* Hon. J. V. Brower, St. Paul, Minn., 1897.

CHAPTER VII

BEAVER TAILS AND WILD HONEY

¹ *Journal of a Tour into the Interior of Missouri and Arkansaw Performed in the Years 1818 and 1819,* Henry R. Schoolcraft, p. 67. London.
² *Missouri Historical Review,* Vol. 21, p. 214.
³ See the Fayette *Advertiser,* October 15, 1869.
⁴ *A History of the Pioneer Families of Missouri,* William S. Bryan and Robert Rose, pp. 107-109. St. Louis, Mo., 1876.

CHAPTER VIII

THE MORMON MIGRATION

¹ See *James Bridger,* J. Cecil Alter, Salt Lake City, 1925.

CHAPTER IX

DIVING FOR FURS

¹ *The Kansas Historical Quarterly,* Vol. II, No. 4, November, 1933, p. 343. Topeka, 1933.
² "Reminiscences of Frederick Chouteau," *Transactions of the Kansas State Historical Society,* 1903-1904, Vol. VIII, p. 425.

Chapter X
HELL ON THE BORDER

[1] See *The Making of Buffalo Bill*, A Study in Heroics, Richard J. Walsh, p. 79, Indianapolis, 1928.
[2] *American Ballads and Folk Songs*, collected and compiled by John A. Lomax and Alan Lomax, pp. 132-133. New York, 1934.

Chapter XI
KAYCEE

[1] Kansas City *Star*, October 1, 1913.
[2] *Kansas City Jazz*, Decca Album #214. Notes by Dave E. Dexter, Jr.
[3] *Ibid.*, p. 5.
[4] *Southwest Review*, Vol. XXIX, No. 1, Autumn, 1943, p. 104.

Chapter XII
AK-SAR-BEN

[1] *Our Singing Country*, A Second Volume of American Ballads and Folk Songs, collected and compiled by John A. Lomax and Alan Lomax; Ruth Crawford Seeger, Music Editor, pp. 233-234. New York, 1941.
[2] Kansas City *Star*, November 22, 1942.
[3] The leather pouch containing the mail. It consisted of a large rectangle of leather covering the saddle like modern saddle skirts, with a padlocked pocket on each corner to hold letters. The rider sat upon it and could quickly remove it and transfer it to a new saddle as often as he changed mounts.
[4] See *Harper's Magazine*, Vol. 177, July and August, 1938.

Chapter XIII
CHIEF BLACKBIRD AND SERGEANT FLOYD

[1] For reproductions in color of some of Bodmer's paintings, see *Life*, Vol. 15, No. 9, pp. 87 ff.

NOTES

[2] *The Trail of Lewis and Clark, 1804-1904,* Olin D. Wheeler, pp. 166-167. New York.
[3] *Audubon's Journal,* ii, 1897, p. 72, at date of July 2, 1843.

CHAPTER XIV

THE PETRIFIED MAN

[1] *Warpath,* Stanley Vestal, p. 94, Boston, 1934.

CHAPTER XV

RANGE AND GRANGE

[1] As quoted by W. R. Felton, of Sioux City.

CHAPTER XVI

MOUNTAIN MEN

[1] *History of the American Fur Trade,* Hiram M. Chittenden, Vol. II, p. 606. New York, 1902.

CHAPTER XVII

STANDING ROCK

[1] For this information and the authentic legend, I am indebted to Mr. Frank Zahn of Fort Yates.
[2] See his amusing book *Sundown LeFlare,* New York, 1899.

CHAPTER XVIII

FOUR BEARS

[1] "Magic and Sleight-of-Hand Performances by the Arikara," George F. Will, *North Dakota Historical Quarterly,* 1928-1930, Vol. III, No. 1, p. 64.

Chapter XIX
CUSTER AND COMANCHE

[1] In those days in the Army a man might hold three ranks at once—one in the regular army, one in the volunteers, and one by brevet. Custer, at the time of his death, was a Lieutenant Colonel in the regular army, a brevet Colonel, and a Major General of volunteers. As usual after a war, men who had been generals dropped back (after the Civil War) to the rank of field officers. In a single military communication with endorsements written by Generals Terry, Sheridan, and Sherman in 1876, the year of his death, Custer is referred to variously as "Lieutenant Colonel Custer" and "Colonel Custer" and "General Custer."

[2] *Boots and Saddles, or Life in Dakota with General Custer*, Elizabeth B. Custer, p. 263. New York, 1885.

[3] *Keogh, Comanche and Custer*, Edward S. Luce, Captain E.O., U.S. Army, Retired.

Chapter XX
GHOST DANCE

[1] *Transactions and Reports of the Nebraska State Historical Society*, Vol. III, pp. 185-190. Fremont, Nebr., 1892.

[2] See *Sitting Bull*, Stanley Vestal. Boston, 1932.

[3] See "The Ghost Dance Religion," James Mooney, p. 865, *Fourteenth Annual Report, Bureau of American Ethnology*, Part 2, Washington, D.C., 1896.

[4] See letter dated March 13, 1917, to Commissioner of Indian Affairs.

[5] "The Last Indian War 1890-91—A Study of Newspaper Jingoism," Elmo Scott Watson. Reprinted from *Journalism Quarterly*, Vol. XX, No. 3, September, 1943.

[6] See "The Ghost Dance Religion," James Mooney, in *Fourteenth Annual Report of the Bureau of American Ethnology*, 1892-93, Part 2, p. 883, Washington, Government Printing Office, 1896.

Chapter XXI
THE MISSOURI RIVER WOMEN

[1] *Mandan and Hidatsa Music*, Frances Densmore. Smithsonian Institution, Bureau of American Ethnology, Bulletin 80, pp. 98-99. Washington, 1923.

NOTES

[2] *The Cheyenne Way, Conflict and Case Law in Primitive Jurisprudence*, K. N. Llewellyn and E. Adamson Hoebel. Norman, University of Oklahoma Press, 1941. By permission of the publishers.

[3] *Sacajawea*, Grace Raymond Hebard, p. 269. Arthur H. Clark Co., Glendale, Cal., 1933.

Chapter XXII
THE LITTLE MISSOURI

[1] *Roosevelt in the Bad Lands*, Hermann Hagedorn, p. 348. Boston and New York, 1921.

Chapter XXIII
FORT UNION

[1] *Journal of Rudolph Friederich Kurz*, Translated by Myrtis Jarrell, Edited by J. N. B. Hewitt, pp. 168-170. Smithsonian Institution, Bureau of American Ethnology, Bulletin 115, Washington, 1937.

Chapter XXIV
THE GREAT DAM

[1] *Congressional Record*, Seventy-eighth Congress, First Session. Remarks of the Hon. Karl E. Mundt of South Dakota in the House of Representatives, Monday, October 4, 1943.

[2] "The Missouri River, Its Habits and Eccentricities Described by a Personal Friend," George Fitch, *The American Magazine*, Vol. LXIII, No. 6, April, 1907, p. 637.

Chapter XXV
HONEYMOON CRUISE

[1] *History of Early Steamboat Navigation on the Missouri River; Life and Adventures of Joseph La Barge*, Hiram Martin Chittenden, (2 vols.), New York, 1903.

Chapter XXVI
WHITE CASTLES

[1] See *The River and I*, John G. Neihardt, pp. 141-142. New York, 1910.

[2] For reproductions in black and white of paintings of the White Castles, etc., see (1) Karl Bodmer's plates, Nos. 18, 67, 68, 70, 73, and 74, illustrating *Travels in the Initerior of North America, 1832-1834*, republished as Vol. XXV of *Early Western Travels* by Reuben Gold Thwaites, Cleveland, 1906; (2) *Life*, Vol. 15, No. 9, August 30, 1943, pp. 87 and 88 for color reproductions.

[3] *History of the Expedition Under the Command of Captains Lewis & Clark*, Nicholas Biddle, Editor, Vol. I, pp. 312-314. New York, 1906.

Chapter XXVII
CHIEF JOSEPH'S LAST BATTLE

[1] *New Northwest*, October 12, 1877.

[2] *Chief Joseph, The Biography of a Great Indian*, Chester Anders Fee, p. 254. New York, 1936.

[3] *Report Sec. of War*, 1877, p. 632.

[4] For Indians' accounts, see *New Sources of Indian History*, and *Sitting Bull*, Stanley Vestal.

Chapter XXVIII
GYPSIES OF THE UPPER RIVER

[1] *North Dakota Historical Quarterly*, Vol. I, No. 1, October, 1926, pp. 41-42 and 44. State Historical Society of North Dakota, Grand Forks and Bismark, N.D.

Chapter XXIX
GREAT FALLS

[1] *The Log Cabin Studio of Charles M. Russell, Montana's Cowboy Artist*, The Russell Memorial Committee, Great Falls, Mont.

NOTES

Chapter XXX

THE GATES OF THE MOUNTAINS

[1] *Indian and White in the Northwest, A History of Catholicity in Montana, 1831 to 1891*, L. B. Palladino, S.J. Lancaster, Pa., 1922.
[2] *Ibid.*
[3] *Ibid.*, p. 189.
[4] *Board of Trade Journal*, April, 1889.

Chapter XXXI

THREE FORKS

[1] These lines, entitled "Others May Praise What They Like," are given here as they first appeared in *Drum-Taps*, 1865.

Acknowledgments

I TOOK my first boat trip on the Missouri River nearly forty years ago and since then have been up and down the stream so often that it is now impossible to remember and record all my obligations to the people along the stream and elsewhere who have contributed to my understanding of it. Trying to put the Missouri River into one volume is like trying to pack an elephant into a sardine can, and limitations of space have sometimes made it impossible to include much interesting material that these friends gave me, but I am nonetheless grateful for their help, which contributed so much to my knowledge of the subject.

I am obligated to those in charge of nearly all the libraries, historical societies, and chambers of commerce along the river, to the Friends of the Middle Border, to the Library of Congress, the Bureau of American Ethnology, and the archives of the Commissioner of Indian Affairs.

I wish to thank here the editors of the *American Magazine* for their kind permission to quote the passages by George Fitch included in the book and to the director of the University of Oklahoma Press for his generous permission to use the passages quoted from *The Cheyenne Way* by Llewellyn and Hoebel (Norman, 1941). I am grateful for the facilities offered me by the curator of the Phillips Collection at the University of Oklahoma. I wish also to express my deep appreciation of the help given by Mr. Paul I. Wellman and the Kansas City *Star*.

Especially I wish to thank Mrs. M. I. Draper of Montana; Mr. Harvey McCaleb of Mission, Kansas; Mr. W. R. Felton of Sioux City, Iowa; Dr. Paul Sears of Oberlin, Ohio; Mr. Lawrence K. Fox of the State Historical Society, Pierre, South Dakota; Mr. Russell Reid of the State Historical Society, Bismarck, North Dakota; Mr. W. A. Falconer of Bismarck, North Dakota; Colonel S. C. Vestal, U.S.A., Retired, of Pasadena, California; Mr. F. H. Rice of Sioux City, Iowa; Professor Lewis E. Atherton of the University of Missouri; Mr. Harry A. Robinson of Yankton, South Dakota; the Reverend Joseph S. Ewing of St. John's Church, Norman, Oklahoma; Mr. Fred H. Monfore of *Yankton Press and Dakotan*, Yankton, South Dakota; Mr. Lew L. Callaway of Helena, Montana; Mr. Frank Zahn, the Upper Missouri Interpreter, of Fort Yates, North Dakota; Mr. George W. Nelson of Northwood, Iowa; Mr. Addison E. Sheldon of the State Historical Society of Nebraska; Mr. John S. Wright of Indianapolis, Indiana; Mr. L. L. Dickerson of Indianapolis, Indiana; Mr. George J. Remsburg of Porterville, California; Mrs. F. G. Frink of Palo Alto, California; Chief White Bull, Circling Hawk, Robert P. Higheagle, Old Bull, Gray Whirlwind (all of the Sioux); and George Peo-peo-tah-likt, John Moses, Philip Andrews, Charley Kow-to-likt (Nez Percés); and Hump, Bob-Tail-Horse (Cheyennes); Mr. J. H. Allen of Yankton, South Dakota; Mr. A. McG. Beede; Captain J. M. Belk; the late General William C. Brown, U.S.A., Retired; Mr. John P. Carignan; the late Mr. Lewis F. Crawford; Mr. Joseph Dietrich; Mr. Melvin R. Gilmore; the late Mr. George Bird Grinnell; Miss Grace Raymond Hebard; the late Dr. V. T. McGillycuddy; Mr. Doane Robinson; the Reverend Father Bernard Strassmeier, O.S.B.; Mr. Frederick Weygold; the late Mr. William Presley Zahn; Mr. Chris Emmett, of San Antonio, Texas; and the late John Homer Seger.

I wish also to thank the following helpful friends: Mrs.

ACKNOWLEDGMENTS

Alma H. Aultman of Miami Beach, Florida; Mrs. Bertha Baker, State Department of History and Archives, Des Moines, Iowa; Mr. Arley R. Bjella of Williston, North Dakota; Miss Betty Boyd of Nebraska City Public Library; Mr. A. J. Breitenstein of Great Falls, Montana; Mr. I. R. Bundy of St. Joseph, Missouri; Miss Era T. Canon of Council Bluffs, Iowa; Mr. E. E. Collins of Vermillion, South Dakota; Mr. Charles H. Compton of St. Louis, Missouri; Mr. Leo C. Dailey of Sioux City, Iowa; Miss Elsie Evans of Leavenworth, Kansas; Mr. Ward R. Evans of Sioux City, Iowa; Miss Louise M. Fernald of Great Falls, Montana; Miss Irene Gentry of Kansas City, Missouri; Miss Iva Glessner of Sioux City, Iowa; Mrs. Gertrude Henderson of Sioux City, Iowa; Mr. R. C. Henry of Helena, Montana; Mr. Emery Hoenshell of Omaha, Nebraska; Mrs. Ethel C. Jacobsen of Pierre, South Dakota; Mr. N. Jay Leonard of Leavenworth, Kansas; Miss Verna Leonard of Plattsmouth, Nebraska; Mr. F. R. Mares of Niobrara, Nebraska; Miss Ethel Martin of the Iowa State Historical Society; Mrs. Anne McDonnell of Helena, Montana; Miss Helen McFarland of the State Historical Society, Topeka, Kansas; Miss Mabel T. Miller of Helena, Montana; Mr. W. H. Over, Director of the Museum of the University of South Dakota, Vermilion, South Dakota; Mr. James A. Painter of Boonville, Missouri; Mr. G. R. Palen of St. Louis, Missouri; Mr. J. L. Rader of the University of Oklahoma Library, Norman, Oklahoma; Mr. R. R. Robinson of Washburn, North Dakota; Mr. M. Schamber of the Mobridge Civic Association, Mobridge, South Dakota; Mrs. Lucinda B. Scott of the Historical Society of Montana; Mrs. Elizabeth Shelly of Atchison, Kansas; Mr. Floyd C. Shoemaker of the State Historical Society of Missouri, Columbia, Missouri; Mr. Thomas Thomsen of Nebraska City, Nebraska; Miss Margaret Eugenia Vinton of Jefferson City, Missouri; and Miss Elizabeth J. Young of Lexington, Missouri.

Bibliography

A COMPREHENSIVE bibliography on the Missouri River would more than fill this book. I have consulted the historical archives, the published collections, and the periodicals of the state and many of the county historical societies of the region, files of the principal newspapers, and some papers in private hands, in addition to government reports and publications, the state guides of those commonwealths touching the river, the *Waterways Journal,* and the *Mississippi Valley Historical Review.* I have contented myself with listing here my principal sources and a number of well-known authors.

AUDUBON, MARIA R., *Audubon and His Journals,* edited by Elliott Coues, New York, 1897.
BIDDLE, NICHOLAS (Editor), *History of the Expedition Under the Command of Captains Lewis and Clark,* Vol. I. New York, 1906.
BRACKENRIDGE, HENRY MARIE, *Journal of a Voyage Up the River Missouri in 1811* (2nd ed., 1816). Also in Vol. 6, *Early Western Travels,* edited by Reuben Gold Thwaites (32 vols., 1904-1906).
BROWER, HON. J. V., *The Missouri River and Its Utmost Source.* St. Paul, Minn., 1897.
BURLINGAME, MERRILL G., *The Montana Frontier.* State Publishing Company, Helena, Mont.
CATLIN, GEORGE, *North American Indians* (two vols.). Being letters and notes on Their Manners, Customs, and Conditions,

Written During Eight Years' Travel Amongst the Wildest Tribes of Indians in North America, 1832-1839. Edinburgh, 1926.

CHAPPELL, PHILIP EDWARD, *A History of the Missouri River*. Kansas City, 1911.

CHITTENDEN, HIRAM MARTIN, *The American Fur Trade of the Far West*. New York, 1902.

────── "List of Steamboat Wrecks on the Missouri River from the Beginning of Steamboat Navigation to the Present Time . . ." being a part of Appendix WW of the *Annual Report of Chief of Engineers for 1897*. Washington Government Printer, 1897. (55th Cong. 2nd Sess. House Doc. 2, pp. 3870-3892.)

────── *History of Early Steamboat Navigation on the Missouri River; Life and Adventures of Joseph La Barge* (2 vols.). New York, 1903.

CHITTENDEN, HIRAM MARTIN, and ALFRED TALBOT RICHARDSON, *Life, Letters and Travelers of Father Pierre-Jean De Smet, S. J., 1801-1873*. Four Volumes. New York, 1905.

CLAYTON, WILLIAM, *William Clayton's Journal*. Published by the Clayton Family Association. Salt Lake City, Utah, 1921.

CONNELLEY, WILLIAM ELSEY, *Quantrill and the Border Wars*. Cedar Rapids, Iowa, 1910.

CRAWFORD, LEWIS F., *Rekindling Camp Fires*. The Exploits of Ben Arnold (Connor). Bismarck, N. D., 1926.

CUSTER, ELIZABETH B., *"Boots and Saddles,"* or Life in Dakota with General Custer. New York, 1885.

DALE, HARRISON CLIFFORD (Editor), *The Ashley-Smith Explorations and the Discovery of a Central Route to the Pacific, 1822-1829*. Cleveland, 1918.

DEATHERAGE, C. P., *Steamboating on the Missouri River in the Sixties*. Kansas City, 1924.

DENSMORE, FRANCES, *Mandan and Hidatsa Music*. Bureau of American Ethnology, Bulletin 30. Washington, 1923.

DE VILLIERS, BARON MARC, *La Découverte du Missouri et l'Histoire du Fort d'Orléans*, (1673-1728).

DIMSDALE, PROF. THOS. J., *The Vigilantes of Montana or Popular Justice in the Rocky Mountains*. Virginia City, Montana, 1921.

DODGE, MAJOR GENERAL GRENVILLE M., Chief Engineer Union

Pacific Railway, 1866-1870, *How We Built the Union Pacific Railway*. Council Bluffs, Iowa.

FAULKNER, EDWARD H., *Plowman's Folly*. University of Oklahoma Press, Norman, Okla. 1943.

FEE, CHESTER ANDERS, *Chief Joseph, The Biography of a Great Indian*. New York, 1936.

FITCH, GEORGE, "The Missouri River, Its Habits and Eccentricities Described by a Personal Friend," *American Magazine*, Vol. LXIII, No. 6, April, 1907.

FLANDRAU (MRS.) GRACE C. (HODGSON), *Frontier Days Along the Upper Missouri*. Great Northern Railway (c. 1927).

FOWKE, GERARD, *Antiquities of Central and Southeastern Missouri*. (Report on Explorations made in 1906-7 Under the Auspices of the Archaeological Institute of America). Smithsonian Institution, Bureau of American Ethnology, Bulletin 37. Washington, 1910.

FREEMAN, LEWIS R., "Trailing History Down the Big Muddy." *The National Geographic Magazine*. Vol. LIV, No. 1. July, 1928.

GARRAGHAN, GILBERT J., *Chapters in Frontier History*. Research Studies in the Making of the West. Milwaukee, 1934.

GASS, PATRICK, *Gass's Journal of the Lewis and Clark Expedition*. Chicago, 1904.

GOLDER, FRANK ALFRED (in Collaboration with Thomas A. Bailey and J. Lyman Smith), *The March of the Mormon Battalion from Council Bluffs to California*. Taken from the Journal of Henry Standage. New York and London.

HAGEDORN, HERMANN, *Roosevelt in the Bad Lands*. Boston and New York, 1921.

HAINES, FRANCIS, *Red Eagles of the Northwest*, the Story of Chief Joseph and his People. Portland, Oregon, 1939.

HANSON, J. M., *Conquest of the Missouri*. Chicago, 1909.

HEBARD, GRACE RAYMOND, *Sacajawea*. Glendale, Calif., 1933.

HEREFORD, ROBERT A., *Old Man River*. The Memories of Captain Louis Rosche, Pioneer Steamboatman. Caldwell, Idaho, 1943.

HICKS, JOHN D., *The Populist Revolt*. A History of the Farmers' Alliance and the People's Party. Minneapolis, 1931.

History of the Expedition Under the Command of Captains Lewis and Clark. A complete reprint of the Biddle Edition of 1814. New York, 1922.

HOWARD, HELEN ADDISON (Assisted in the Research by Dan L. McGrath), *War Chief Joseph.* Caldwell, Idaho, 1941.

KURZ, RUDOLPH FRIEDERICH, *Journal of Rudolph Friederich Kurz.* Translated by Myrtis Jarrell, Edited by J. N. B. Hewitt. Smithsonian Institution, Bureau of American Ethnology, Bulletin 115. Washington, 1937.

LARPENTEUR, CHARLES (Edited by Elliott Coues), *Forty Years a Fur Trader on the Upper Missouri.* The Personal Narrative of Charles Larpenteur. New York, 1898.

LEWIS, MERIWETHER, *History of the Expedition Under the Command of Lewis and Clark* (Edited by Elliott Coues). New York, 1893.

———, *Original Journals of the Lewis and Clark Expedition, 1804-06* (8 vols.). Printed from the original manuscripts in the library of the American Philosophical Society. New York, 1904-05. Vol. I includes a bibliography.

LLEWELLYN, K. N., and ADAMSON, HOEBEL E., *The Cheyenne Way.* Conflict and Case Law in Primitive Jurisprudence. Norman, University of Oklahoma Press, 1941.

Log Cabin Studio of Charles M. Russell, Montana's Cowboy Artist. Published and copyrighted by The Russell Memorial Committee, Great Falls, Mont.

LOMAX, JOHN A. and ALAN (Collected and compiled by), *American Ballads and Folk Songs.* New York, 1934.

LUCE, EDWARD S. (Captain, E.O., U.S. Army, Retired), *Keogh, Comanche and Custer.* St. Louis: Privately published, 5356 Page Boulevard, 1939.

MARGRY, PIERRE, *Découvertes et établissements des français dans l'ouest et dans le sud de l'Amérique Septentrionale (1614-1754). Mémoires et documents originaux recuellis et pub. par.* (6 vols.). P. Margry, Paris, 1879-81.

MCDONALD, W. J., "The Missouri River and Its Victims; Vessels Wrecked from the Beginning of Navigation to 1925," *Missouri Historical Review,* XXI, January, 1927, pp. 215-242; April, 1927, pp. 455-480; July, 1927, pp. 581-607.

BIBLIOGRAPHY

McWhorter, Lucullus Virgil, *Yellow Wolf: His Own Story.* Caldwell, Idaho, 1940.

Mooney, James, *The Ghost Dance Religion and the Sioux Outbreak of 1890.* Fourteenth Annual Report, Bureau of American Ethnology, Part 2, Washington, D.C., 1896.

Mott, Frank Luther, *American Journalism, 1690 to 1940.* New York, 1941.

Murray, Hon. Charles Augustus, *Travels in North America During the Years 1834, 1835 and 1836* (2 vols.). London, 1839.

Neihardt, John Gneisenau, *The River and I.* New York, 1910.

Russell, Charles M., *Trails Plowed Under.* Garden City, N. Y., 1927.

———, *Good Medicine;* Memories of the Real West. Garden City, N. Y. First published, 1929.

Schoolcraft, Henry R., *Journal of a Tour into the Interior of Missouri and Arkansaw Performed in the Years 1818 and 1819.* London.

Smith, Charles Edward (with Frederic Ramsey, Jr., Charles Payne Rogers, and William Russell), *The Jazz Record Book.* New York, 1942.

Smith, Winston O., *The Sharps Rifle, Its History, Development and Operation.* New York, 1943.

Tabeau, Pierre A., *Tabeau's Narrative of Loisel's Expedition to the Upper Missouri.* Translated from the French by Rose Abel Wright. Norman, University of Oklahoma Press, 1939.

Thwaites, Reuben Gold (Editor), *Early Western Travels, 1748-1846.* Cleveland, Ohio, 1906.

——— (Editor), *Original Journals of the Lewis and Clark Expedition, 1804-1806* (8 vols., 1904-05).

———, *Travels in the Interior of North America, 1832-1834.* Republished as Vol. XXV of *Early Western Travels.* Comprising the series of original paintings by Charles Bodmer. Cleveland, 1906.

Vestal, Stanley, *Mountain Men.* Boston, 1937.

———, *New Sources of Indian History, 1850-1891.* University of Oklahoma Press, Norman, Oklahoma, 1934.

Vestal, Stanley, *Sitting Bull, Champion of the Sioux*. A Biography. Boston and New York, 1932.

Villard, Oswald Garrison, *John Brown, 1800-1859. A Biography Fifty Years After*. Boston and New York, 1911.

Violette, Eugene Morrow, *A History of Missouri*. New York, 1918.

Wheeler, Olin D., *The Trail of Lewis and Clark, 1804-1904* (2 vols.). New York and London, 1904.

Where These Rocky Bluffs Meet, Including the Story of the Kansas City Ten-Year Plan. Published by the Chamber of Commerce of Kansas City, Mo., 1938.

Whitney, Mrs. C. W., *Kansas City, Missouri, Its History and Its People, 1808-1908* (3 vols.). Chicago, 1908.

Wied-Neuwied, Maximilian [Alexander Philipp], prinz Von, *Reise in das innere Nord-America.* . . . Coblenz, 1839-1841. London edition, translated into English, 1843: Maximilian, Prince von Wied's *Travels in the Interior of North America, 1832-1834*. Edited by Reuben Gold Thwaites and reprinted as volumes XXII-XXV of *Early Western Travels, 1748-1846*, Arthur H. Clark Co., Cleveland, O., 1906.

Index

Abolition, 110.
Abolitionists, 107, 110, 112.
Achilles, 162.
Ak-sar-ben, 136, 137.
American Fur Company, 324.
Anderson, Bill, 115, 116.
Anderson, Pilot, P. J., 286.
Arbor Lodge, 137.
Army, 102.
Ashley, Maj. William H., 184-195.
Astorians, 18, 32.
Atchison, Senator David, 111.
Audubon, John James, 140, 148, 272.

Badlands, 232, 261, 263-265, 294-296.
Baker house, 60.
Baptiste—*see* Pomp.
Barada, 157.
Barker, Dr., 104.
Basie, Count, 127.
"Battle Hymn of the Republic," the, quoted, 110.
BATTLES: Battle of the Badlands, 265; Battle of the Big Hole, 307; Battle of Killdeer Mountain, 265; Battle of the Washita, 219; Battle of Westport, 115; Crooked River Fight, 91; Grattan Fight, mentioned, 223; Haun's Mill Fight, 91.
Bear's Heart, 265.
Becknell, William, 80.
Beecher, Henry Ward, quoted, 112.
"Beecher's Bibles," 112.
Belcourt, quoted, 314-315.
Belknap, William K., 220.
Belt, Capt. Francis T., 46.
BENDS: Alert, 42; Augusta, 43; Bonhomme, 43; Box Car, 43; Brickhouse, 43; Bushwhacker, 43; Cora, 43; Disaster, 285; Great, 42, 43; Malta, 43; Nigger, 43; Pelican, 43; Sheep Nose, 43; Tabo, 43.
Bent, William, 124.
Benteen, Gen. Frederick W., 223.
Big Basin, 94.
Big Foot, 235, 236, 238.
Bingham, George Caleb, 15, 117.
Bird Woman, 257.

Black robes, 323.
Blackbird, Omaha chief, 140-145.
Boat Woman, 257.
Bodmer, Karl, 56, 144, 213, 293.
Boggs, Gov. of Mo., 91.
Bois de Boulogne, 64.
Book of Mormon, the, 85, 86.
Boone, Daniel, 68, 73, 75, 76.
Border Ruffians, 113, 114.
Boys' Town, 136.
Brackenridge, H. M., 19, 24, 27, 29, 31, 32, 103.
Bradbury, John, 19, 32.
Bridger, Jim, 96, 184.
Brooke, Gen. John P., 231, 236.
Brown, John, 105, 113-115.
Buffalo Bill, 115, 124, 162, 233 ff., 234.
Bull Head (Cheyenne), 253-255.
Bullboat, the, 158 ff.
Bullhead (Sioux), 234.
Burkman, Pvt. "Old Nutriment," 225.
"Bushwhacking," 24.
Butterfield Overland Mail Line, 131, 133.

Caldwell County, Mo., 90.
Calf-Woman, 155, 156.
Caliban, 12.
Canada, 57, 67, 113, 173, 177, 183, 290, 301, 304, 310.

Carafel, 273.
Carson, Kit, 81, 82, 124, 184.
Carson, Moses B., 192, 195.
Casanova, 273.
Catch-The-Bear, 234.
Catholics, 109.
Catlin, George, 144, 148, 206, 213, 322.
Cats on the Missouri R., 103.
Chappell, Capt. Phillip E., quoted, 51.
Charbonneau, Toussaint, 246-253, 257.
Chardon, F. A., 276, 277.
Cheyenne River Reservation, 235.
Chicagou, 64, 65.
Chief Joseph, 299-310, 311.
Chouteau, 272.
Chouteau, Francois G., 99.
Chouteau, Frederick, 99, 100, 101.
Chouteau, Pierre, 55, 270.
Christian missionaries, 322-325.
Church of Jesus Christ of Latter Day Saints, 86.
Clark, Gen. John B., 91.
Clark, William, 28; quoted, 146, 322.
Clay County, Mo., 89.
Clayton, William, composer, 95.
Clendennin, Colonel, 302.
Cleveland, Pres. Grover, 230.
Clubb, Henry S., 105.
Coast, the, 131.

INDEX

Cody, Col. William F. — see Buffalo Bill.
Colter, John, 184, 331.
Comanche, cavalry mount, 224 ff.
Confederacy, 60.
Congress, 68, 71, 76, 94, 106, 109, 280.
Cooper, Milly, 78, 79.
Coronado, Francisco Vásquez, 136.
Cowboy, the national hero, 163 ff.
Coyote (mythical trickster), 157.
Craft, Father, 239.
Culbertson, Alexander, 272.
Custer, Mrs. Elizabeth B., 222.
Custer, Gen. George Armstrong, 103, 124, 219-225, 308.
Custer, Col. Tom, 198 ff.
Cutter, Dr. Calvin, 112.

"Dad" Lemmon, 162.
Danites, the, 90.
Datchurut, Baptiste, 100.
De Bourgmond, Etienne Veniard, 61, 62, 63.
De Mores, Marquis, 261-262.
De Smet, Father Pierre-Jean, 51, 323-325.
De Villiers, Baron Marc, 61.
Democrats, 91, 230.
Denig, Edwin T., fur trader, 148, 274.

Dixon, Billy, 124.
Dodge, Gen. Grenville, 134.
Don Juan, 273.
Donald Duck, 157.
Doniphan, Alex W., 88.
Dorion, Pierre, 19.
Dougherty, John, 100.
Draper, Mrs. M. I., 291.
Du Bois, Sergeant, 64.
Dunklin, Gov. of Mo., 88.
Dycke, Professor Lewis Lindsay, 225.

Earp, Wyatt, 124.
Eleventh Kansas Infantry Volunteer Regiment, 116.
English, the, 67.
Ewing, Gen. Thomas, 116.

Falstaff, Sir John, 57.
Febold Feboldson, 157.
Femme Osage Section, 76.
Fetterman, Capt. W. J., 223.
Fifth Infantry, 304.
Fireheart, Sioux chief, 192, 193.
Fitch, George, quoted, 8, 52, 284.
Fitzpatrick, Tom, 184, 192.
Flanagan, Father, 137.
Floyd, Sgt. Charles, 145-147.
Foley, Michael, 302.
Ford brothers, Bob and Charles, 120, 121, 122.

INDEX

Ford, Gov. of Ill., 94.
Forsyth, Col. James W., 236, 239.
Fort Peck Dam, 280-284, 293.
FORTS: Abraham Lincoln, 60, 198, 201, 219; Assiniboin, 290; Atkinson, 190; Belknap, 229; Benton, 47, 51, 131, 285 ff., 299, 301, 302; Berthold, 213, 229, 270, 278, 279; Buford, 131, 226, 288, 311; Cantonment Martin, 55; Clark, 270; Cooper's Fort, 78; D'Orleans, 62, 63, 65; Dilts, 261; Hempstead, 78; Keogh, 301; Lapwai, 300; Laramie, 223; Leavenworth, 131; Mandan, 42, 204, 247; McKenzie, 270, 276, 277; Missouri, 55; "Old Fort," 59; Osage, 28, 55, 185; Peck, 229; Phil Kearny, 223; Pierre, 270; Randall, 226; Recovery, 192; Rice, 324; Sumter, 106, 115; Tecumseh, 55; Union, 131, 148, 267-277, 285; Washakie, 258; Yates, 196 ff.
Four Bears, 209 ff.
France, 61, 62, 63, 64, 66, 68, 107; King of, 71.
Franklin, Benjamin, 163.
Free Silver, 110.
Free Soilers, 114.
Freebooters, 114.
Fulton, Robert, 51.

Gall, Sioux chief, 198, 287-288.
Gallagher, Jack, 124.
Garion, 275.
"Garry Owen," 221.
Gerard, 278-279.
Ghost Dance, 110, 226-242.
Ghost shirts, 230 ff.
Gibbon, Gen. John, 222, 299.
Gibson, Reed, wounded, 189.
Glass, Hugh, 184, 189.
Good Medicine, 321.
Gordon, William, 192.
Grant, Pres. Ulysses S., 220.
Grass, John, 198.
"Grasshoppering," 45, 49.
Gray Eyes, Ree chief, 193.
Gray Whirlwind, 217, 218.
Gré, 273-276.
Great American Desert, 109.
Great Britain, 67, 71.
Great Plains, the, 131, 162, 163.
Great Salt Lake, the, 96.
Guépe, 272.

Hailstorm, 170.
Hale, Capt. Owen, 305, 307.
Hamilton, James Archdale, 271.
Handbook of American Indians North of Mexico, 69.
Hanson, Capt. J. N., cited, 50.
Harney, Gen. William S., 324.
Harvey, Alexander, 276-277.
Helen of Troy, 327.
Henry, Maj. Andrew, 33, 184-195.

INDEX 359

Henry, Maj. Guy V., 240.
Hercules, 12.
Hollady, Overland Stage Line, 131.
Horn, William, 150.
Howard, Gen. O. O., 300 ff.
Howard, Thomas—*see* Jesse James.
Hudson's Bay, 12.
Hudson's Bay Company, 177, 183.
Hughes, 100.
Hunt, Wilson P., 18-33 *passim.*, 98.

Ice, 44.
Iktomi, 157.
Ilges, Maj. Guido, 301, 302.
Indian opinions on Ghost Dance, 231-232.
INDIANS: Algonquin, 70, 71; Arikara—*see* Rees; Assiniboin, 272; Bannocks, 301; Blackfeet, 71, 157, 212, 270, 273, 322, 324, 331; Blood, 276, 277; Bois Brulés—*see* Métis; Cheyenne, 71, 203, 236, 304; Chippewa, 312; Comanche, 257-258; Cree, 312; Crow, 158, 299, 301, 331; Delaware, 86; Flatheads, 71, 322, 323, 324, 331; Fox, 69; Gros Ventres of the Village—*see* Hidatsa; Hidatsa, 204, 212, 247, 256, 270; Illinois, 63, 69; Kaw, 100; Kite, 158; Mandans, 204, 212, 214, 242-246, 279; Métis, 311-315; Minnetarees *see* Hidatsa; Missouri, 61, 63, 69, 134, 145; Nez Percés, 299-310, 311, 322, 323, 331; Omaha, 30, 141-145; Osages, 28, 61, 63, 214; Oto, 63, 69, 134, 145; Padouca, 62; Pawnees, 71; Red River Halfbreeds—*see* Métis; Rees, 32, 71, 185-195, 212, 214, 270; Sac, 69; Shawnee, 86, 104; Shoshone—*see* Snakes; Sioux, 7, 18, 31, 69, 140, 156, 157, 191 ff., 217, 218, 226-242, 270, 304, 306, 310, 311, Hunkpapa, 226, Minniconjou, 155; Yankton, 158; Snakes, 71, 247, 248, 252, 253, 331.
Iron Horse, 200 ff.
Irving, Washington, quoted, 75, 186.
ISLANDS: Bon Homme, 150; Cedar, 27; Cedar (No. 2), 31; Cow, 13, 42, 54; Cow (Mont.), 290, 298, 301, 302, 303; Snake, 98; Wizard's, 27.
Isle of Orleans, 67.

Jackson County, Mo., 86, 87, 88, 89, 97, 116, 120.
Jackson, Gen. Andrew, 53, 76.
James, Jesse, 117-122, 123, 135.

Janey, 257, 258.
Jayhawkers, 114.
Jefferson, Pres. Thomas, 68, 71, 109, 145, 319, 330.
Jessaume, 2 47, 248.
Jesse James Hotel, 122.
Joutel, 69.
Judith Basin, 320.

Kansans, 109, 111.
Kansas City jazz, 127.
Kansas-Nebraska Act, the, 107, 108.
Kaufman, Louie, 320.
Keane, Charles, 13.
KEELBOATS: described 22, 23; *Yellowstone Packet*, 185, 190; Rocky Mountains, 185, 190.

Kelley, newshawk, 239.
Keogh, Capt. Myles W., 224.
King Louis XV, 63.
King of the Missouri—*see* Kenneth McKenzie.
Know-nothings, 109.
Kurz, Rudolph Friedrich, 15, 51, 273.

La Barge, Capt. Joseph, 51, 291.
La Salle, Nicholas, 69.
La Salle, Robert, 69.
Lamanites, 86, 96.
LANDMARKS: Arrow Rock, 43; Bird Beak Peak (Eagle's Nose), 242 ff., Blackbird Hill, 30, 139, 140, 145; Blacksnake Hills, 42; Boon's Lick, 75; Boon's Lick Spring, 76; Brulé Bottom, 42, 277, 285; Burned Butte, 292; Cathedral Rock, 292; Center Monument, the, 106; Citadel Bluff, 292; Coalbanks, the, 42, 290; Côte Sans Dessein, 27; Council Bluff (treaty ground), 55, 134; Cow Creek Canyon, 299, 302; Cow Island Crossing, 299, 310; Devil's Race Ground, 42; Eagle Rock, 292; Elk Horn Prairie, 42; Floyd's Bluff, 30, 146, 147; Fort Rock, 330; Four Bears Bridge, 213; Gates of the Mountains, 328, 329, 330; Giant Spring, 319; Grand Canyon of the Little Missouri, 264; Great Falls, 130, 250, 299, 316-321, 331; Great Sulphur Spring, 250; Haystack Butte, 292; Hole in the Wall, 42; Manitou Rocks, 27; Mule's Head Landing, 42; Musick Ferry, 42; Osage Chute, 43, 54; Painted Woods, 42; Peru Cutoff, 42; Point Labadie, 25; Pompey's Pillar, 256; Pomp's Tower, 256; Portage La Force, 42; Prime's Ferry, 100; Proposal Hill, 197 ff.; Ramparts, the, 42; Rattlesnake

INDEX

Springs, 42; Snake Butte (S. D.), 106; Snake Bluffs, 98; Split Rock, 27; Standing Rock, 42, 196 ff., 226, 229; Stone Walls, 42, 297-298, 316; Square Butte, 42; Tavern Rock, 24; White Castles, 294.
Lane, Senator James H., 114, 116.
Larkin, Col. Thomas B., 283.
Le Borgne, Hidatsa chief, 247.
Leavenworth, Col. Henry, 190 ff.
Lee, Rev. Daniel, 323.
Lee, Rev. Jason, 323.
Lee, Robert E., 115.
Leighton, George R., 135.
Lewis and Clark, 15, 57, 62, 68, 69, 71, 98, 103, 134, 140, 144, 145, 151, 158, 161, 185, 197, 204, 246-253, 255-258, 292, 297, 299, 316-319, 330-331.
Lewis, Caption Meriwether, 147, 315-319; quoted, 328, 329.
Lincoln, Abraham, 52, 113, 133, 134; quoted, 108, 111.
Lindbergh, Charles, 285.
Lisa, Manuel, 16, 18-33 *passim.*, 98, 186, 331.
Little Assiniboin, 293.
Little Bow, 142.
Little River Women Society, 246.

Livingston, Edward, American ambassador, 68.
Lomax, John A., quoted, 128.
Long, Maj. Stephen H., 53, 54, 55.
Looking Glass, Chief, 304 ff.
Lost tribes of Israel, 160, 204.
Louisiana Purchase, 107, 319.
Lower River, 55, 73, 75, 161, 185, 284.
Luce, Capt. Edward S., 225.
Lulu, Joseph, 101.

MacDonald, Angus, 192.
Mackinaw, 15, 16, 303, 310.
Magical performances, 214 ff.
Mandan earth lodge, described, 205 ff.
Mandan Okeepa, described, 206-209.
Mark Twain, 35, 131.
Marquette, Father, 11, 69.
Marsh, Capt. Grant, 223, 291.
Mato-Tope—*see* Four Bears.
Maynadier, Lt. H. E., quoted, 207-209.
McGillycuddy, Dr. V. T., Indian Agent, 230 ff.
McKenzie, Kenneth, 212, 270-272.
McLaughlin, Major James, Indian Agent, 197, 233, 234.
Messiah—*see* Wovoka.
Methodist Episcopal Church, 323.

INDEX

Mexicans, 111.
Mickey Mouse, 157.
Miles, Gen. Nelson A., 233, 301-310; quoted, 238.
MINING CAMPS: Alder Gulch, 325; Bannack, 325; Confederate Gulch, 325; Diamond City, 327; Last Chance Gulch, 325; Montana Bar, 326; Virginia City, 325.
Mint, the, 319, 321.
Miro, Spanish governor, quoted, 67.
Missouri Compromise, 106.
Missouri Fur Company, 18, 19, 191, 331.
Missouri Legion, 192 ff.
Missouri meerschaum, 75.
MISSOURI RIVER (names for), Anati (Hidatsa), 214; Big Muddy, 5, 7, 18, 48, 66, 97, 103, 106, 129, 131, 226, 242, 281, 284, 315; Mata (Mandan), 214; Muddy Water, 70; Pekitanoui, 11; River St. Phillip, 71; Smoky River, 30; Wooden Canoe, 14; Etymology of name, 69-72.
Missouri State University, 80.
Missourians, 87, 88, 109.
Mitchell, Col. David D., 274.
Moelchert, Sgt. William, 301.
Moncrevier, Jean Baptiste, 275.
Monroe, James, Minister Extraordinary, 68.

Montana State Guide Book, 292.
Moore, Ely, quoted, 99.
Mormon battalion, the, 95.
Mormonism, 110.
Mormons, 46, 85-97, 111, 117.
Morton, J. Sterling, 137.
Moses, Gov. John, 174.
Mountain Men, 179-195.
MOUNTAINS: Bear Paw, 303; Big Belt, 325, 328; Bitter Root, 322; Black Hills, 176, 219-220, 259; Little Rockies, 303; Rocky, 6, 33, 53, 131, 166, 226, 247, 323, 331; Shining, 6.
Mowgli, 152.
Murray, the Hon. Charles Augustus, quoted, 74.

Napoleon Bonaparte, 68, 71.
Nauvoo Legion, 94.
Nelson, Captain, 53.
Nelson, William Rockhill, 124.
New England, 110.
New England Emigrant Aid Company, 108.
New Deal, 280.
NEWSPAPERS: *Frontier Guardian*, the, quoted, 96; *Kansas City Star*, 124, 125, 126; *Morning and Evening Star*, Mormon paper, 88; *Missouri Intelligencer*, quoted, 53, 80; *Missouri Republican*, the,

184; *St. Louis Inquirer*, quoted, 54.
No-Horns-On-His-Head, 322.
Non-Partisan League, 110, 166.
North Dakota Plan, 173.
Northwest passage, 54.
Notre Dame de Paris, 64.

Odysseus, 242, 246.
O'Fallon, Maj., Indian Agent, 191.
Old Man, 157.
Order Number Eleven, 116, 117.
Ordinance of 1787, 68.
O'Reilly, Don Alessandro, Spanish governor, 67.
"Others May Praise What They Like," quoted, 332.

Pacific Ocean, 53, 54.
Packinaud, 273.
Palladino, Father L. B., 323-325.
Parker, Rev. Samuel, 323.
Paul Bunyan, 157.
Paul, Prince of Württemberg, 257.
Pharaoh, 167.
Pilcher, Joshua, 186-195.
Pine Ridge, 229, 230, 232, 235.
Plains Indians, 183.
Plowman's Folly, 175.
Plummer, Henry, 327.
Pomp, 248, 250, 251, 253, 256, 257.
Pony Express, 132, 133.
Populism, 110.
Prather, Pvt. W. H., (9th Cav.), ballad quoted, 240-242.
Price, Gen. Sterling, 116.
Princess of the Missouri, 59, 64.
Prohibition, 110, 126.

Quantrill, William Clarke, 115, 116.
Quivera, Kingdom of, 136, 137.
Quixote, Don, 27.

Rabbit-Skin-Leggins, 322.
Rain-In-The-Face, 198 ff.
Raynolds, Gen. W. F., 207.
Rector, Col. Elias, 53.
Red Cloud, Sioux chief, 231.
Red Legs, 115.
Red Queen and Alice, 48
Remington, Frederic, 202 ff., 321.
Reno, Maj. Marcus A., 223, 224.
Republicans, 230.
Rice, William G,. 77.
Riel, Louis, 315.
Rigdon, Sidney, 86, 90.
Roi, Baptiste Louis, 79, 80.
Romulus, 152.
Roosevelt, Theodore, 261-263.
Rose, Edward, 186, 187, 192.
Rosebud Agency, 229.

Rough Riders, 263.
Royal Northwest Mounted Police, 299.
Royer, D. F., Indian Agent, 230, 231.
Running Antelope, 198.
Russell, Charlie, 166, 319-321.
Russell, Majors, and Waddell, 131.

Sacajawea, 246 ff., 248-253, 255-258, 331.
St. James Hotel, 123.
Santa Claus, 165.
Schoolcraft, Henry R., quoted, 74, 75, 212.
Schreyvogel, Charles, 321.
Second Cavalry, 304, 306.
Seventh Cavalry, 198 ff., 219 ff., 235, 304, 305.
Seventh Infantry, 301.
Shanghai Pierce, 123.
Sharp Nose, 255.
Sharps rifle, 112.
Shawnee Baptist Mission, 104.
Sheridan, Gen. Phillip Henry, 220.
Siren song, 246.
Sirens, 242.
Sitting Bull, 103, 161, 162, 198, 220, 226 ff., 233, 234, 235, 265, 293, 301, 306, 310, 311, 324.
Slade, J. A., 327.
Smallpox epidemic, 211, 213.

Smith, Hyrum, 94.
Smith, Jedediah, 184, 190 ff.
Smith, Joseph, 85, 86, 87, 90, 91, 94.
Snags, 44.
Songs of the Rivers of America, 128.
Southwest Review, the, 128.
Spain, 66, 67, 68.
Spaniards, 67.
Speed, Joshua F., 108.
Standing Rock Agency, 196 ff., 233.
STATES: Arkansas, 94, 117; California, 94, 95, 104, 133, 325; Colorado, 325; Idaho, 300; Illinois, 91, 94; Iowa, 6, 95, 97, 133, 166, 280; Kansas, 6, 13, 106, 107, 108, 109, 111, 112, 113, 114, 117, 166, 174, 280; Kentucky, 53, 67, 73, 76, 80, 86, 147; Louisiana, 58; Minnesota, 166; Missouri, 6, 13, 88, 89, 106, 107, 109, 114, 115, 117; Montana, 6, 177, 278, 280, 324, 325, 326, 327; Nebraska, 6, 96, 97, 107, 133, 136, 166, 280; New York, 86; North Dakota, 6, 99, 166, 173, 174, 259, 280; Ohio, 53, 86, 88, 89; Oklahoma, 117, 177, 236, 257, 310; Oregon, 94, 103, 300, 323; South Dakota, 6, 166, 261, 270, 280; Tennes-

see, 67, 73, 76, 86; Texas, 67, 94, 177; Virginia, 76, 80; Wyoming, 258, 280.
Steamboat (personal Indian name), 56.
STEAMBOATS: 33, 34-58 *passim.*, 86; *A. B. Chambers*, 35; *Arabia*, 112, *Assiniboin*, 56; *Belle of St. Louis*, 39; *Benton*, 286, 301, 303; *Chippewa*, 285-286; *Expedition*, 54; *Far West*, 224, 286; *Florence*, 286; *Fontanelle*, 286; *General Mead*, 286 ff.; *Ida Stockdale*, 285; *Independence*, 53; *Long's Dragon*, 50-58 *passim*.; *Luella*, 285; *Mary Mac-Donald*; 286; *Montana*, 39; *Moses Greenwood*, 50; *Nellie Peck*, 286, 289; *Only Chance*, 286; *R. M. Johnson*, 54; *St. Ange*, 51; *Saluda*, 46; *Thomas Jefferson*, 54; *Welcome*, 286; *Western*, 286; *Western Engineer*, 54; *Yellowstone*, 55, 56, 98.
STREAMS (Not Tributaries): Big Horn R., 224; Cherry Creek, 151; Clearwater, the, 300; Colorado R., 280; Columbia R., 33, 54, 250, 280; Congo, the, 149; Gold Creek, 325; Grasshopper Creek, 325; Green R., 324; Little Big Horn R., 202, 221, 223;

Powder R., 324; Rhine, the, 144, 292; Snake Creek, 304; Willamette R., 323; Wounded Knee Creek, 235, 239; Zambesi, the, 149.
STREAMS (Tributaries): Bad (Teton) R., 55, 259, 270; Big Blue R., 88; Bird Woman's R., 250; Burnt Creek, 278-280; Cheyenne R. ("the Fork"), 43, 151, 155, 191, 229; Crooked Creek, 256; Gallatin R., 330; Gasconade R., 27; Grand R. (Mo.), 61, 98; Grand R. (S. D.), 71, 185, 226, 229; Heart R., 219; Jefferson R., 330; Kaw R., 28, 54, 61, 98, 99, 103, 105, 106, 123; Little Missouri R., 261 ff.; Little Missouri R. (so-called), 55, 259; Little Platte R., 28; Madison R. 330; Marias R., 316; Milk R., 303, 311; Mississippi R., 11, 48, 66, 67, 94; Musselshell R., 293, 301, 302; Nodaway R., 28; Osage R., 27; Platte R., 61, 69, 128; Shell R., 250; Slaughter R., 297; Stone-Idol Creek, 197; Sun R., 317; Teton R.—*see* Bad R.; Thick Timber R., 264; Three Forks, 166, 247, 330-331; Vermilion Creek, 30; Yellowstone R., 12, 22, 138,

185, 190, 221, 224, 226, 256, 288, 301.
Stringfellow, 109.
Sturgis, Col. Samuel, 299 ff., 307.
Sublette, William, 192.
Sully, Gen. Alfred, 264, 265.
Sun Dance, 178, 229.
Sutton, Bill, 149.
Switzerland, 292.

Tennessee Valley, 280.
TERRITORIES: Dakota, 57; Indian, 117, 310; Louisiana, 67, 68, 71.
Terry, Gen. Alfred H., 220.
"The Caissons Go Rolling Along," 128.
"The Illinois," 66.
"The Wide Missouri," 128.
Théâtre Italien, 64.
Tonty, 69.
Too-hul-hul-suit, 300, 309.
TOWNS: Atchison, 54, 107, 131; Bismarck, 131, 206, 228, 286; Bluffton, 39; Boonville, 73; Carrollton, 60; Chariton, 53; Chicago, 177; Columbia, 80, 91; Côte Sans Dessein, 27, 79; Council Bluffs, 95, 131, 133, 134, 135, 140; Deadwood (S. D.), 259; Elbowoods, 259; Far West, 90, 91; Forest City, 149, 150; Fort Benton, 320, 326; Fort Peck, 280, 281, 292; Franklin, 53, 80, 81, 85, 131; Gallatin City, 331; Great Falls (Mont.), 319; Harpers Ferry (Va.), 114; Hart's Bluff, 95; Helena (Mont.), 320, 327, 328; Independence, 86, 87, 89, 97, 99, 100, 131; Kanesville, 95; Kansas City, 5, 13, 39, 57, 99, 104 108, 123, 284; Kickapoo, 107; Lawrence, 108, 111, 113; Leavenworth, 54, 107, 131; Les Petites Côtes, 69; Lexington, 27, 46, 112; Liberty, 90; Macy, 144; Mandan (N. D.), 57, 131; Marmarth, 261; Medora, 261 ff.; Miami, 59; Miller's Hollow, 95; Nauvoo, 94, 95; New Franklin, 81; New Orleans, 61, 63, 66, 68; Omaha, 131, 133, 135, 136; Omaha Village, 30; Paris, 63, 64, 66; Pierre (S. D.), 106, 131; Platte City, 13; Ponca Village, 30, 31; Quebec, 66, 69; Sacramento, 133; Santa Fe, 35, 81, 86, 103; St. Charles, 22, 68, 69, 75, 76, 131; St. Joseph, 5, 39, 62, 120, 122, 131, 133; St. Louis, 5, 50, 53, 55, 72, 86, 99, 141, 148, 274, 285, 322, 324; St. Paul, 177; Sioux City, 57, 131, 147; Three Forks, 331;

INDEX

Tipton, 131; Vermilion, 287; Washington, D. C., 317, 330; Washington (Mo.), 75; Waverly, 60, 62; Weston, 50; Westport, 104, 131, 324; Winter Quarters, 96; Yankton, 131.

TRAILS: Boon's Lick, 76, 80; Mexican Trace—*see* Santa Fe Trail; Oregon, 5; Santa Fe, 5, 80, 177.

Trails Plowed Under, 321.

TREATIES: of Ghent, 79; of Laramie, 324; of Paris, 67; of San Ildefonso, 68.

Ulysses, 162.
Underground railroad, 107.
Union Army, 60.
United States, 68.
United States Treasury, 280.
Upper Louisiana, 66, 67, 76.
Upper River, 29, 55, 144, 160, 174, 176, 178, 267, 270, 278, 294, 315, 321; varied population, 178 ff.

Van Hocken, Father, 51.
Vancouver Island, 94.
Vanderburgh, Henry, 192, 195.
Vasquez, Mrs. Barnett, 100.
Vegetarian Kansas Emigration Company, the, 104.
Verendrye, Sieur de la, 204.

Verse quoted, 8, 26, 32, 95, 110, 116, 117, 121, 122, 129, 130, 132, 240-242 246, 259-261, 265-266, 327, 332.
Vigilantes, 326 ff.
Vilandre, 273-276.
Von Wied, Prince Maximilian, 55, 98, 144, 209, 276, 292, 293.

Waddell, William Bradford ("Bible Bill"), 131, 132, 133.
Wallowa Valley, 300.
Warriors' Society, the, 158.
WARS: Civil, 39, 55, 60, 106, 134, 219; French and Indian, 67; Ghost Dance, 231, 240; Mexican, 88, 95; Spanish American, 263; War of 1812, 78, 79.
Washinga Sahba—*see* Blackbird.
Washington, George, 163.
Whigs, 91.
Whistling Bear, 279.
White Bird, 300, 310, 311.
White Bull, Sioux chief, 152-155, 265.
White Cow, Indian chief, 148.
White Savages, 160.
White, William Allen, quoted, 125.
Whitman, Walt, quoted, 331-332.
Whitman, Dr. Marcus, 323.

Wi-Jun-Jon (the Pigeon's Egg Head), 148.
Wild Bill (Hickok), 124, 259-261.
Wisconsin Historical Society, 147.
Woman - Who - Lived - With - Wolves, 152-155.
Women suffrage, 110.
Wongatap, 209 ff.
Woolfolk, Capt. Charles P., 286 ff.
Woolfolk, Mrs. Charles P., 286-291.

Wordsworth, William, quoted, 135.
World's Columbian Exposition, 149.
Wovoka, 228 ff.

Yellow Bird, 237.
Yellowstone Park, 331.
York, Clark's servant, 251.
Young, Brigham, 65, 94, 95, 96.

Zabette, Frank, 100.

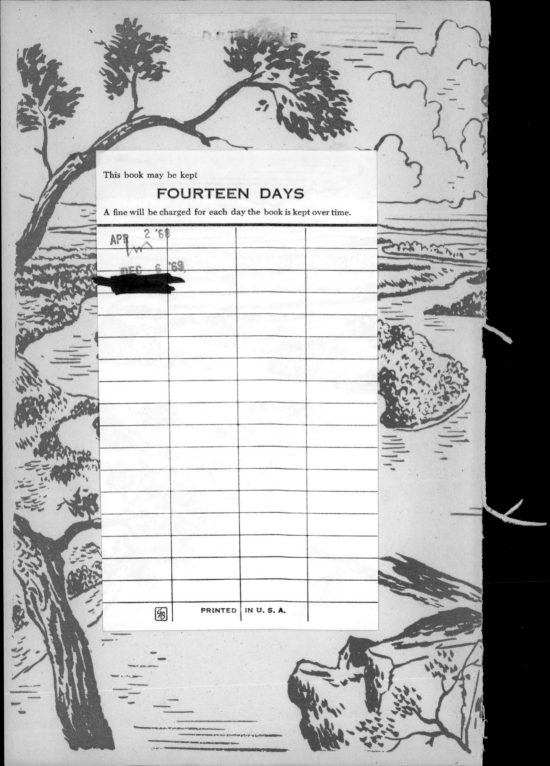